SpringerBriefs in Space Developn

Series Editor

Joseph N. Pelton

For further volumes:
http://www.springer.com/series/10058

This Springer book is published in collaboration with the International Space University. At its central campus in Strasbourg, France, and at various locations around the world, the ISU provides graduate-level training to the future leaders of the global space community. The university offers a two-month Space Studies Program, a five-week Southern Hemisphere Program, a one-year Executive MBA and a one-year Masters program related to space science, space engineering, systems engineering, space policy and law, business and management, and space and society.

These programs give international graduate students and young space professionals the opportunity to learn while solving complex problems in an intercultural environment. Since its founding in 1987, the International Space University has graduated more than 3,000 students from 100 countries, creating an international network of professionals and leaders. ISU faculty and lecturers from around the world have published hundreds of books and articles on space exploration, applications, science and development.

Siamak Khorram · Frank H. Koch
Cynthia F. van der Wiele
Stacy A. C. Nelson

Remote Sensing

INTERNATIONAL®
SPACE UNIVERSITY

Springer

Siamak Khorram
Department of Environmental Science,
 Policy, and Management
University of California
Berkeley, CA 94720
USA

and

Center for Earth Observation
North Carolina State University
5123 Jordan Hall 7106
Raleigh, NC 27695-7106
USA

Frank H. Koch
Forestry Sciences Laboratory
USDA Forest Service
Southern Research Station
3041 E. Cornwallis Road
Research Triangle Park, NC 27709
USA

Cynthia F. van der Wiele
715 Shepherd St.
Durham, NC 27701
USA

Stacy A. C. Nelson
Center for Earth Observation
North Carolina State University
5123 Jordan Hall 7106
Raleigh, NC 27695-7106
USA

ISSN 2191-8171
ISBN 978-1-4614-3102-2
DOI 10.1007/978-1-4614-3103-9
Springer New York Heidelberg Dordrecht London

e-ISSN 2191-818X
e-ISBN 978-1-4614-3103-9

Library of Congress Control Number: 2012930489

Printed on acid-free paper

Springer is part of Springer Science+Business Media (www.springer.com)

Contents

Chapter 1
Introduction

We humans have long been interested in documenting our world. For example, what is arguably the earliest surviving map was engraved on a stone block approximately 14,000 years ago (Utrilla et al. 2009). Found in a cave in the Spanish Pyrenees, the block appears to show features from the surrounding landscape—mountains, rivers, and lakes—as well as potential access routes to different parts of the landscape. The researchers who discovered the block proposed that it might be a planning map for an upcoming hunt, or a depiction of one that had already occurred. Although the field of cartography (i.e., mapmaking) obviously made huge strides in the intervening millennia, it was not until the twentieth century A.D.—after the invention of the airplane and the refinement of cameras and photographic techniques—that humans had the capacity to observe a sizeable geographic area at one time, and just as importantly, to record these "remotely sensed" observations for further study and comparison. Moreover, starting with the launch of the first Earth-observing satellites in the 1950s and 1960s, we have been able to examine our world and its changes from a truly global perspective.

Today, a vast array of data-collecting instruments, mounted on both satellites and airplanes, permit scientists to measure our influence on a range of Earth systems, enabling researchers, for example, to track rates of deforestation, ice melt from glaciers and polar areas, trails of air pollution, the extent of sea-level rise, and many other regional and global phenomena. These instruments, along with techniques to analyze the data they collect, are constantly improving, which not only makes them more sensitive to change, but also more reliable sources of information for decision-makers. The objective of this book is to provide you with a basic introduction to remote sensing and showcase a variety of its applications in fields ranging from traditional earth sciences to natural resources, engineering, physical geography, and the social sciences and humanities.

S. Khorram et al., *Remote Sensing*, SpringerBriefs in Space Development,
DOI: 10.1007/978-1-4614-3103-9_1, © Siamak Khorram 2012

What is Remote Sensing and Why Do It?

Remote sensing is defined as the acquisition and measurement of information about certain properties of phenomena, objects, or materials by a recording device not in physical contact with the features under surveillance. This is a rather broad definition that encompasses, for instance, medical technologies such as X-rays and magnetic resonance imaging (MRI). In an environmental context, remote sensing typically refers to technologies for recording **electromagnetic energy** that emanates from areas or objects on (or in) the Earth's land surface, oceans, or atmosphere (Short 2010). Essentially, the properties of these objects or areas, in terms of their associated levels of electromagnetic energy, provide a way to identify, delineate, and distinguish between them. Because the electromagnetic properties of these features are commonly collected by instruments mounted on aircraft or Earth-orbiting spacecraft, remote sensing also gives scientists the opportunity to capture large geographic areas with a single observation, or **scene** (Fig. 1.1).

Another potential advantage of remote sensing, especially when done from satellites, is that geographic areas of interest are revisited on a regular cycle, facilitating the acquisition of data that reveals changing conditions in these areas over time. For a given instrument, or **sensor**, onboard a satellite, the revisit time depends on the satellite's orbit as well as the width of the sensor's **swath**, which is the track the sensor observes as its satellite travels around Earth. This concept of a sensor's **temporal resolution** will be explored further in Chap. 2 of this book. Remotely sensed data are **geospatial** in nature, meaning that the observed areas and objects are referenced according to their geographic location in a geographic coordinate system, such that they may be located on a map (Short 2010). This allows the remotely sensed data to be analyzed in conjunction with other geospatial data sets, such as data depicting road networks or human population density. Indeed, remotely sensed data with sufficient detail can be used to characterize such things as the growth or health status of vegetation, or to identify habitat edges or ecotones (i.e., transition zones between ecological communities) that could not otherwise be discerned effectively from maps created from field observations (Kerr and Ostrovsky 2003).

This point illustrates the unique importance of remote sensing as a data source for **geographic information systems (GIS)**, which are organized collections of computer hardware, software, geographic data and personnel designed to efficiently capture, store, update, manipulate, and analyze all forms of geographically referenced information (Jensen 2005; ESRI 2001). In turn, **geographic information science** is concerned with conceptual and scientific issues that arise from the use of GIS, or more broadly, with various forms of geographic information (Longley et al. 2001). Terms such as **geomatics** and **geoinformatics** have similar meanings. Ultimately, the problem-solving value of a GIS depends upon the quality of the data it contains (Jensen 2005; Longley et al. 2001).

Fig. 1.1 In remote sensing, an instrument (i.e., sensor or scanner) is mounted on an aircraft or satellite that records information about objects or areas on the ground. Typically, the data documents the amount of electromagnetic energy associated with the target. The extent, or footprint, of the geographic area captured depends on the sensor's design and the altitude of the craft in which it is mounted

The Electromagnetic Spectrum

Electromagnetic radiation (EMR) is defined as all energy that moves with the velocity of light in a **harmonic wave pattern** (i.e., all waves are equally and repetitively spaced in time). Visible light is just one category of EMR; other types include radio waves, infrared, and gamma rays. Together, all of these types comprise the **electromagnetic spectrum** (Fig. 1.2). As illustrated by Fig. 1.2, the different forms of EMR vary across the spectrum in terms of both **wavelength** and **frequency**. Wavelength is the distance between one position in a wave cycle to the same position in the next wave, while frequency is the number of wave cycles passing the same point in a given time period (1 cycle per s = 1 Hertz, or Hz). The mathematical relationship between wavelength and frequency is expressed by the following equation: $C = \lambda \cdot v$, where λ is wavelength, v is frequency, and C is the speed of light (which is constant at 300,000 km per s in a vacuum). Visible light, representing only a small portion of the electromagnetic spectrum, ranges in wavelength from about 3.9×10^{-7} (violet) to 7.5×10^{-7} m (red), and has

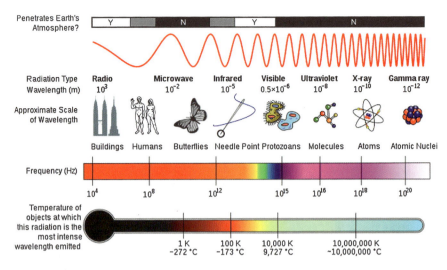

Fig. 1.2 The electromagnetic spectrum. Illustration courtesy of NASA

corresponding frequencies that range from 7.9×10^{14} to 4×10^{14} Hz (Fig. 1.3). (Note that EMR wavelengths are commonly expressed in nanometers, where 1 nm $= 10^{-9}$ m, or micrometers, where 1 μm $= 10^{-6}$ m.)

When EMR comes into contact with matter (i.e., any object or material, such as trees, water, or atmospheric gases), the following interactions are possible: **absorption**, **reflection**, **scattering**, or **emission** of EMR by the matter, or **transmission** of EMR through the matter. Remote sensing is primarily based on detecting and recording reflected and emitted EMR. Fundamentally, what makes remote sensing possible is that every object or material has particular emission and/or reflectance properties, collectively known as its **spectral signature** or **profile**, which distinguishes it from other objects and materials. In turn, remote sensors are attuned to collect these "spectral" data.

A remote sensor records these data in either analog (i.e., **aerial photographs** collected with an aircraft-mounted film camera) or, more commonly, digital format (i.e., a two-dimensional matrix, or **image**, composed of pixels that store EMR values recorded by a satellite-mounted array) (Jensen 2005). Furthermore, sensors may be either **passive** or **active** in nature. Passive sensors—the predominate category of sensors currently operated around the world—record naturally occurring EMR that is either reflected or emitted from areas and objects of interest. In contrast, active sensors—such as microwave (i.e., RAdio Detection And Ranging, or **radar**) systems—send manmade EMR toward the features of interest and then record how much of that EMR is reflected back to the system (Jensen 2005). Chap. 2 of this book provides details about many contemporary remote sensing systems, both active and passive.

Fig. 1.3 "Boulevard du Temple, Paris," a photograph (daguerreotype) taken by Louis Daguerre in 1838

Photo Interpretation, Photogrammetry and Image Processing

Prior to the widespread availability of satellite imagery, aerial photography served as the principal foundation for a wide variety of cartographic efforts and geographic analyses (Short 2010; also see the next section of this chapter). **Air photo interpretation** uses characteristics such as tone, texture, pattern, shadow, shape, size, and site (location), to identify objects and areas in photos, while **photogrammetry** is the science or art of obtaining reliable spatial measurements of objects from their aerial photographs. (The word "photogrammetry" is derived from three Greek roots meaning "light-writing-measurement.") The types of measurements that a photogrammetrist might collect include the distances between features in an area of interest, the areal extent, or the heights of particular features. In recent decades, photogrammetric methods have been increasingly applied to digital rather than film-based aerial photographs, and in some cases, to imagery captured by satellite-based sensors (e.g., Konecny et al. 1987). However, this book focuses primarily on non-photo-interpretive and non-photogrammetric applications of remotely sensed images. In particular, Chap. 3 discusses **digital image processing**, which is a concept that encompasses a wide variety of computer algorithms and approaches for visualization, enhancement, and interpretation of remotely sensed images. Key products of digital image processing include thematic maps and color-coded, classified images that depict the spatial pattern of certain characteristics of objects and features. For instance, remote sensing is commonly applied to map the variety and extent of land cover; in turn, these maps serve as critical information for other applications such as ecological studies, urban planning, water resource monitoring, and environmental impact assessment, as well as for policy making (Hester et al. 2008, Khorram et al. 1996, Kerr and Ostrovsky 2003, Franklin and Wulder 2002).

The Importance of Accuracy Assessment

A key consideration regarding the use of remotely sensed data is that the resulting outputs must be evaluated appropriately. When the output is a classified map, it is critical to assess its accuracy. A map is an imperfect representation of the phenomena it is meant to portray. In other words, every map contains errors, and it is the responsibility of the remote sensing analyst to characterize these errors prior to a map's use in subsequent applications. The most widely accepted method for the accuracy assessment of remote-sensing-derived maps is by comparison to reference data (also known as "ground truth") collected by visiting an adequate number of sample sites in the field (Congalton and Green 1999; Khorram et al. 1999). The key instrument in this comparison is an error matrix, which quantifies the accuracy for each map class of interest as well as the overall map accuracy (i.e., combining all of the classes). This technique will be discussed in greater detail in Chap. 3 of this book.

A Brief History of Remote Sensing

One of the key advances that laid the groundwork for the field of remote sensing was the invention of photography, which enabled the near-instantaneous documentation of objects and events. The French inventor Joseph Niépce is generally credited with producing the first permanent photograph in 1826, which showed the view from his upstairs workroom window (Hirsch 2008). In 1839, it was announced that Louis Daguerre—who collaborated with Niépce until his death in 1833—had invented a process for creating a fixed silver image on a copper plate, which he called a daguerreotype (Newhall 1982). One of his daguerreotypes, "Boulevard du Temple, Paris" (Fig. 1.3), taken in 1838 or 1839, is reputedly the oldest surviving photograph of a person. The image appears to shows an empty street, but this is an artifact; due to the relatively long exposure time (more than ten minutes), the image failed to capture moving carriage and pedestrian traffic. However, Daguerre was able to capture a man who stopped to have his shoes shined (see lower left of Fig. 1.3).

Notably, Daguerre's "Boulevard Du Temple" image has many characteristics of what is now called an **oblique** aerial photograph (i.e., an aerial photograph captured from an angle rather than vertically, or directly overhead) (Mattison 2008). In any case, the potential cartographic applications of photography were almost immediately recognized.

In 1840, François Arago, the director of the French Académie des Sciences and the man who publicly announced Daguerre's process, advocated the use of photographs to produce topographic maps (Mattison 2008, Wood 1997). Nevertheless, credit for the first actual aerial photograph is given to the French photographer Gaspar Felix Tournachon, who used the pseudonym "Nadar." Nadar patented the concept of using aerial photography for cartography and surveying in 1855, but experimented unsuccessfully until 1858, when he captured a photograph from a balloon tethered 80 m above the Bievre Valley (PAPA International 2011). Unfortunately, none of Nadar's early efforts are believed to have survived. The oldest existing aerial photograph is a view of Boston, taken from a balloon by James Wallace Black in 1860 (Fig. 1.4).

During the latter part of the nineteenth century and into the early twentieth century, a number of people experimented with the use of aerial photography from balloons, kites, and even birds as an effective means of mapmaking and surveying. In 1889, the Canadian Dominion Lands Surveyor General, E.G.D. Deville, published *Photographic Surveying*, a seminal work that focused on balloon-based photography (Mattison 2008). In 1908, Julius Neubronner, a German apothecary and inventor, patented a breast-mounted aerial camera for carrier pigeons (PAPA International 2011). The lightweight camera took automatic exposures at 30-s intervals. Although faster than balloons, the pigeons did not always follow their expected flight paths. When the birds were introduced at the 1909 Dresden International Photographic Exhibition, postcards created from aerial photographs taken by the pigeons were popular with the public (PAPA International 2011). Camera-equipped pigeons were also used for military surveillance. Figure 1.5 shows two examples of pigeon aerial photographs, as well as an image of a pigeon mounted with one of Neubronner's aerial cameras.

In 1880, George Eastman patented a machine for rapidly preparing a large number of "dry" photographic plates (i.e., glass plates coated with a gelatin emulsion). Searching for a lighter and less temperamental alternative to glass plates, he developed rolled paper film, but found paper was not an ideal film base because the resulting photographs tended to be grainy and have inadequate contrast (Utterback 1995). Eastman addressed these limitations with his introduction of flexible celluloid film in 1887. In 1900, Eastman's company, Kodak, released the Brownie, an inexpensive box camera for rolled film, making photography accessible to a mass audience.

Eastman's innovations shortly preceded the Wright Brothers' first successful flight, in 1903, of a heavier-than-air aircraft. Six years later, Wilbur Wright took the first aerial photograph from an airplane (PAPA International 2011). Quickly embraced by military forces, airplane-based aerial photography was used extensively for reconnaissance during the First World War (Rees 2001). Near the end of

Fig. 1.4 The oldest surviving aerial photograph, an image of Boston taken from a balloon in 1860

the war, U. S. entrepreneur Sherman Fairchild began to develop what became the first true aerial camera system (PAPA International 2011). In 1921, Fairchild demonstrated the utility of his system for cartography, employing more than 100 overlapping aerial images to create a **photo-mosaic** of New York City's Manhattan Island (Fig. 1.6). During the period between the First and Second World Wars, aerial photography was also applied in other civilian contexts, including forestry, geology, and agriculture (Rees 2001).

Aerial photography underwent dramatic refinement during the Second World War, a period that also saw the introduction of the first infrared-sensitive instruments and radar imaging systems (Rees 2001). In fact, the basic elements of aerial photography as we know it today were largely formed out of these wartime developments and related technological advances during the next two decades. For example, false-color infrared film was first developed during the Second World

Fig. 1.5 Examples of Julius Neubronner's pigeon aerial photography. Notably, the photograph of the Schlosshotel Kronberg (*top left*) accidentally included the pigeon's wingtips. The image on the *right* shows a pigeon with one of Neubronner's breast-mounted cameras

War for camouflage detection, but was already being applied for air-photo-based mapping of vegetation by the 1950s (Rees 2001).

However, the broader field of remote sensing effectively started with the Space Age, an era initiated by the Soviet Union's launch of the first manmade satellite, Sputnik-1, in 1957. The term "remote sensing" was coined in the mid-1950s by Evelyn Pruitt, a geographer with the U. S. Office of Naval Research, allegedly because the term "aerial photography" did not sufficiently accommodate the notion of images from space (Short 2010). After the launch of Sputnik-1, the U. S. and Soviet governments raced to design and implement new space-related technologies, including both manned spacecraft and satellites.

Although the first (and rather crude) satellite image of Earth was captured by NASA's Explorer 6 in 1959, the U. S. Department of Defense's CORONA (also known as "Discoverer") Reconnaissance Satellite Program, which remained classified until 1995, may be seen as a key forerunner to future Earth-observing satellite programs. During its period of operation (1958–1972), the CORONA Program developed an increasingly sophisticated series of high-resolution, film-based camera systems (Cloud 2001). The first photo of Soviet territory from space, taken in August 1960, shows an air base at Mys Shmidta, Siberia (Fig. 1.7). Within a decade, CORONA satellites had extensively mapped the United States and other parts of the world. Before the CORONA program ended, its science team had begun to experiment with color (i.e., spectral) photography, thus serving as a precursor to the sensors used by the Landsat program (discussed below and in Chap. 2) and modern satellite imaging systems (Cloud 2001).

In the 1960s and into the early 1970s, the United States and Soviet Union launched an assortment of reconnaissance, meteorological, and communications

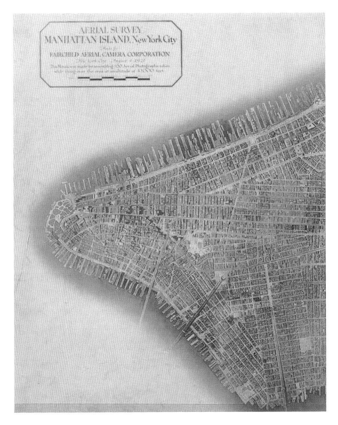

Fig. 1.6 Detail from Fairchild Aerial Camera Corporation's 1921 photo-mosaic of Manhattan, New York, showing the island's southern portion. Image courtesy of Library of Congress, Geography and Map Division

satellites into orbit. Also during this period, astronauts from NASA's Mercury, Gemini, and Apollo space missions took thousands of photographs of Earth using handheld and automated cameras (Witze 2007). In fact, "The Blue Marble," a photograph taken in December 1972 by the crew of Apollo 17, is often cited as the most widely reproduced image of Earth (McCarthy 2009). (In 2002, NASA released a new version of "The Blue Marble," a mosaic of images from the Moderate Resolution Imaging Spectroradiometer (MODIS) instrument onboard the Earth Observing System (EOS) Terra satellite; see Fig. 1.8.)

A more formative event for modern remote sensing occurred in July 1972, when NASA launched *ERTS-A* (Earth Resources Technology Satellite-Mission A), the first satellite dedicated to monitoring environmental conditions on Earth's surface. Shortly after its launch, the satellite's name was changed to ERTS-1. It was followed by ERTS-2 (launched in January 1975) and ERTS-3 (launched in March 1978). Later, the names for these satellites were changed to Landsat-1, -2,

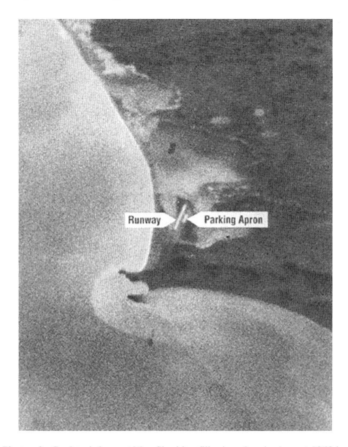

Fig. 1.7 Photo of a Soviet air base at Mys Shmidta, Siberia, taken in August 1960 by a camera onboard the CORONA satellite *Discoverer-14*. Image courtesy of the U. S. National Reconnaissance Office

and -3, respectively. The Landsat program (along with a few other U. S. satellite programs; see Chap. 2) served as the primary source of space-based Earth imagery until the 1980s, when a number of other countries began to develop their own Earth-observing satellite programs, particularly France and the European Union, Canada, Japan, India, Russia, China, and Brazil. More recently, a number of private companies have emerged as providers of satellite imagery, demonstrating the feasibility of commercial, space-based remote sensing.

The European Space Agency (ESA) estimates that, between the 1957 launch of Sputnik and January 1, 2008, approximately 5,600 satellites were launched into Earth orbit (ESA 2009). The vast majority of these are no longer in service (raising some concerns about the high volume of space debris encircling the planet; see Fig. 1.9). Today, just fewer than 1,000 operational satellites are orbiting Earth, approximately 9% of which are dedicated to Earth observation and remote sensing (plus roughly 4% for meteorology and related applications) (UCS 2011). These

Fig. 1.8 A new version of "The Blue Marble": this true-color image is a seamless mosaic of separate images, largely recorded with the Moderate Resolution Imaging Spectroradiometer (MODIS), a device mounted on NASA's Terra satellite. Image courtesy of NASA Goddard Space Flight Center

satellites, carrying a wide range of sensors optimized for various applications, represent a rich potential source of current data for remote sensing analysts. Furthermore, they also extend the nearly four-decade-old satellite image record started by Landsat-1.

A primary goal of this book is to provide readers with a basic understanding of the principles and technologies behind remotely sensed data and their applications. Although it includes some technical information, the book is not really a "how-to" guide for image processing and analysis. Instead, it is our hope that readers will walk away with the ability to confidently use remotely sensed data in their work or research, and to think critically about key questions that users typically must answer:

- Have remotely sensed data been applied to my topic area in the past? In what ways? Could applications from other areas or disciplines inform my current project?

Fig. 1.9 Debris objects in low-Earth orbit. These debris are largely comprised of inactive satellites and other hardware, as well as fragments of spacecraft that have broken up over time. (Objects are not to scale.) Image courtesy of the European Space Agency

- What kind of remotely sensed data do I need? More specifically, what types of data are available from today's remote sensing instruments, and what are their strengths and limitations?
- How must my chosen data be prepared prior to analysis? What are the appropriate processing and/or analytical methods?
- What is the accuracy of the output products I have created? Is that accuracy sufficient for my ultimate objectives?
- What is the possibility that better remotely sensed data will become available in the near future? Realistically, what potential gains am I likely to see from these future data?

References

J. Cloud, Hidden in plain sight: the CORONA reconnaissance satellite programme and clandestine Cold War science. Ann. Sci. **58**, 203–209 (2001)

R.G. Congalton, K. Green, *Assessing the Accuracy of Remotely Sensed Data: Principles and Practices.* (Boca Raton, Lewis Publishers 1999), p. 137

Environmental Systems Research Institute (ESRI) *The ESRI Press Dictionary of GIS Terminology* (Redlands, Environmental Systems Research Institute, 2001), p. 116

European Space Agency (ESA), Space debris: evolution in pictures [web site]. European Space Agency, European Space Operations Centre (2009), http://www.esa.int/esaMI/ESOC/SEMN2VM5NDF_mg_1.html

S.E. Franklin, M.A. Wulder, Remote sensing methods in medium spatial resolution satellite data land cover classification of large areas. Prog. Phys. Geogr. **26**, 173–205 (2002)

D.B. Hester, H.I. Cakir, S.A.C. Nelson, S. Khorram, Per-pixel classification of high spatial resolution satellite imagery for urban land-cover mapping. Photogramm. Eng. Remote Sens. **74**, 463–471 (2008)

R. Hirsch, *Seizing the light: A Social History of Photography*, 2nd edn. (McGraw-Hill Higher Education, New York, 2008), p. 480

J.R. Jensen, *Introductory Digital Image Processing*, 3rd edn. (Prentice Hall, Upper Saddle River, 2005), p. 316

J.T. Kerr, M. Ostrovsky, From space to species: ecological applications for remote sensing. Trends Ecol. Evol. **18**, 299–305 (2003)

S. Khorram, H. Cheshire, X. Dai, J. Morisette, Land cover inventory and change detection of coastal North Carolina using Landsat Thematic Mapper data. In: Proceedings of the ASPRS/ACSM Annual Convention and Exposition, April 22–25, 1996, Baltimore, MD, vol 1: Remote Sensing and Photogrammetry, pp. 245–250

S. Khorram, G.S. Biging, N.R. Chrisman, D.R. Colby, R.G. Congalton, J.E. Dobson, R.L. Ferguson, M.F. Goodchild, J.R. Jensen, T.H. Mace, *Accuracy Assessment of Remote Sensing-Derived Change Detection* (American Society of Photogrammetry and Remote Sensing Monograph, Bethesda, 1999), p. 64

G. Konecny, P. Lohmann, H. Engel, E. Kruck, Evaluation of SPOT imagery on analytical photogrammetric instruments. Photogramm. Eng. Remote Sens. **53**, 1223–1230 (1987)

P.A. Longley, M.F. Goodchild, D.J. Maguire, D.W. Rhind, *Geographic Information Systems and Science* (Wiley, Chichester, 2001), p. 454

D. Mattison, Aerial photography, in *Encyclopedia of Nineteenth-century Photography*, vol. 1, ed. by J. Hannavy (Routledge, New York, 2008), pp. 12–15

J.J. McCarthy, Reflections on: our planet and its life, origins, and futures. Science **326**, 1646–1655 (2009)

B. Newhall, *The History of Photography: from 1839 to the Present*, 5th edn. (Museum of Modern Art, New York, 1982), p. 319

Professional Aerial Photographers Association (PAPA) International. History of aerial photography [web site]. PAPA International (2011), http://www.papainternational.org/history.asp

G. Rees, *Physical Principles of Remote Sensing* (Cambridge University Press, Cambridge, 2001), p. 343

N.M. Short, The Remote Sensing Tutorial [web site]. National Aeronautics and Space Administration (NASA), Goddard Space Flight Center (2010), http://rst.gsfc.nasa.gov/

Union of Concerned Scientists (UCS) UCS Satellite Database [web site]. Union of Concerned Scientists, Nuclear Weapons & Global Security (2011), http://www.ucsusa.org/nuclear_weapons_and_global_security/space_weapons/technical_issues/ucs-satellite-database.html

P. Utrilla, C. Mazo, M.C. Sopena, M. Martínez-Bea, R. Domingo, A palaeolithic map from 13,660 calBP: engraved stone blocks from the Late Magdalenian in Abauntz Cave (Navarra, Spain). J. Hum. Evol. **57**, 99–111 (2009)

J.M. Utterback, Developing technologies: the eastman Kodak story. McKinsey Q. **1**, 130–144 (1995)

Witze, A. News: Briefing—A timeline of Earth observation [online document] Nature (2007), doi:10.1038/news.2007.320. http://www.nature.com/news/2007/071205/full/news.2007.320.html

R.D. Wood, A state pension for L.J.M. Daguerre for the secret of his Daguerreotype technique. Annals of Science **54**, 489–506 (1997)

Relevant Web Sites

The First Photograph (documents the history behind Joseph Niépce's 1826 image, currently housed in the collection of the Harry Ransom Center at the University of Texas): http://www.hrc.utexas.edu/exhibitions/permanent/wfp/

Professional Aerial Photographers Association (PAPA) International—History of Aerial Photography: http://www.papainternational.org/history.asp

Aerial Photographs and Remote Sensing Images—Library of Congress Geography and Maps: An Illustrated Guide: http://www.loc.gov/rr/geogmap/guide/gmillapa.html

Project CORONA: Clandestine roots of modern Earth science—University of California, Santa Barbara: http://www.geog.ucsb.edu/ ~ kclarke/Corona/Corona.html

Landsat Program History: http://landsat.gsfc.nasa.gov/about/history.html

EOEdu, educational website for satellite Earth observation (in French, Dutch, and English): http://eoedu.belspo.be/index.html

Chapter 2
Data Acquisition

In this chapter, we introduce the concept of **resolution** in remote sensing. One common definition of resolution is the ability to discern individual objects or features in a captured image or in the "real world." However, the term also encompasses several specific characteristics of remotely sensed data. We illustrate these specific resolution characteristics with examples of commonly used satellite sensors and imagery. We conclude the chapter with an overview of the types of sensors available to today's analysts, including details about the data they acquire and their potential applications.

Data Resolution

When we talk about "remotely sensed data," we are usually referring to digital images captured by sensors mounted on aircraft or spacecraft. These data are primarily described by four types of resolution: **spatial**, **spectral**, **temporal**, and **radiometric resolution**. **Spatial resolution** is a measure of the fineness of detail of an image. For digital images, this refers to the ground area captured by a single pixel; because pixels are typically square, resolution is generally expressed as the side length of a pixel. **Spectral resolution**, represented by the width of the wavelength interval and/or number of spectral channels (or **bands**) captured by a sensor, defines the storage of recorded electromagnetic energy and the sensor's ability to detect wavelength differences between objects or areas of interest. The amount of time it takes a sensor to revisit a particular geographic location is referred to as its **temporal resolution**. Finally, the sensitivity of a sensor to brightness values (i.e., the smallest differences in intensity that it can detect) is known as its **radiometric resolution**. This metric is usually articulated in terms of binary bit-depth, which refers to number of grayscale levels at which data are recorded by a particular sensor (Jensen 2005). The binary bit-depth is typically expressed in the following ranges of grayscale levels: 8-bit (0–255), 10-bit (0–1,023), 11-bit (0–2,047), 12-bit (0–4,095) and 16-bit (0–65,535).

S. Khorram et al., *Remote Sensing*, SpringerBriefs in Space Development,
DOI: 10.1007/978-1-4614-3103-9_2, © Siamak Khorram 2012

Each of these resolution types is described in greater detail below. But first, the different resolution characteristics of some currently operational satellite sensors are presented in Table 2.1. The sensors described in this table are highly variable with respect to resolution. For example, the Quickbird satellite's onboard sensor has a spatial resolution of 65 cm for **panchromatic** (or black-and-white) imagery, while data collected by the "VEGETATION" sensors on the SPOT 4 and 5 satellites are stored in 1,150-m pixels. Temporally, the sensors listed in Table 2.1 have the capacity to revisit a particular location on Earth's surface every 15 min (i.e., GOES satellites) to every 35 days (i.e., MERIS and other sensors on the Envisat-1 satellite).

The ability to discern spatial structure is an important element of any remote sensing analysis. For instance, in an urban landscape, surface features such as roads, office buildings, parks, and residential neighborhoods comprise a sort of mosaic, where many small constituent units are interspersed with a few large ones. Roads and buildings are typically among the smallest of these units. When viewed from above (e.g., from an airplane), the net visual effect is an aggregate "patch-work" of various land uses and cover types, but the degree of patchwork detail portrayed by a remotely sensed image, and thus the level of specificity at which it can be classified, depends on its spatial resolution (Fig. 2.1). For example, a 30-m pixel (i.e., a spatial resolution provided by several sensors listed in Table 2.1) stores one digital number per spectral band of information for any landscape feature smaller than 900 m^2. The pixel could, in fact, accommodate six one-story square houses, each having a 1500 ft^2 (139 m^2) footprint. In contrast, an Advanced Very High Resolution Radiometer (AVHRR) pixel, with its 1100-m resolution (see Table 2.1), could incorporate more than 8,700 of these houses, if arranged regularly. Increases in spatial resolution have been a persistent trend in sensor innovation. The latest commercial satellites (e.g., GeoEye-1 and WorldView-2) provide spatial resolutions of 1 m or less.

The spectral component of remote sensing data is also critical to appropriate data selection. The spectral resolution of a sensor is a description of the range and partitioning of the electromagnetic energy (or **reflectance**) recorded by the sensor. In this context, sensors can be divided into one of three loose classes of spectral sensitivity: panchromatic (one spectral band), **multispectral** (multiple spectral bands), or **hyperspectral** (many, possibly hundreds, of spectral bands) (Jensen 2005). Fundamentally, both multispectral and hyperspectral images can be thought of as "stacks" of identically geo-referenced (i.e., covering the exact same geographic area) panchromatic images, where each image in the stack corresponds to a particular portion of the electromagnetic spectrum. Conventionally, satellite-derived remote sensing data are captured in panchromatic and/or multispectral format. (Spaceborne hyperspectral imaging, as discussed below, is a relatively recent development). For example, the ETM+ sensor onboard NASA's Landsat-7 satellite records data in eight bands, including one panchromatic band. Each of these bands is sensitive to different wavelengths of visible and infrared radiation. The sensor on the Quickbird satellite records data in four spectral bands targeted at the blue, green, red, and near-infrared portions of the electromagnetic spectrum,

Table 2.1 Characteristics of selected satellite sensors (adapted from Rogan and Chen 2004; updated for this publication)

Sensor (Mission)[a]	Organization[a]	Operation period	Swath width (km)	Spatial resolution (m)[b]	Temporal resolution	Radiometric resolution	Spectral resolution (μm)	Spectral bands
MSS (Landsat 1-5)	NASA, USA	1972–1992	185	80 (MS), 240 (TIR)[c]	16–18 days	8-bit	0.5–1.1, 10.4–12.6[c]	4–5[c]
TM (Landsat 4, 5)	NASA, USA	1982–	185	30 (MS), 120 (TIR)	16 days	8-bit	0.45–2.35, 10.4–12.5	7
ETM+ (Landsat 7)	NASA, USA	1999–	185	15 (PAN), 30 (MS), 60 (TIR)	16 days	8-bit	0.52–0.9 (PAN), 0.45–2.35, 10.4–12.5	7 + PAN
MODIS (EOS Terra and Aqua)	NASA, USA	1999–	2300	250 (PAN), 500 (VNIR), 1000 (SWIR)	1–2 days	12-bit	0.620–2.155, 3.66–14.385	36
ASTER (EOS Terra)	NASA,USA and METI, Japan	1999–	60	15 (VNIR), 30 (SWIR), 90 (TIR)	4–16 days	8-bit (VNIR/SWIR), 12-bit (TIR)	0.52–0.86, 1.60–2.43, 8.125–11.65	14
Hyperion (EO-1)	NASA, USA	2000–	7.5	30	16 days	12-bit	0.353–2.577	220
ALI (EO-1)	NASA, USA	2000–	37	10 (PAN), 30 (MS)	16 days	12-bit	0.48–0.69 (PAN), 0.433–2.35	9 + PAN
CALIOP (CALIPSO)	NASA, USA and CNES, France	2006–	0.1	333	16 days	22-bit[d]	0.532, 1.064	2[c]
AVHRR (NOAA 6-19)	NOAA, USA	1978–[f]	2700	1100	12 h	10-bit	0.58–12.5	6[g]
I-M Imager (GOES 11-15)	NESDIS, USA	1975–[h]	8	1000 (VNIR), 4000 (SWIR), 8000 (moisture), 4000 (TIR)	0.25–3 h	10-bit	0.55–12.5	5
SAR (RADARSAT-1)	CSA, Canada	1995–	45–500[i]	8–100[i]	24 days		Radar	N/A
SAR (RADARSAT-2)	CSA, Canada	2007–	20–500[i]	3–100[i]	24 days[j]		Radar	N/A
MERIS (Envisat-1)	ESA	2002–	1150	300 (land), 1200 (ocean)	35 days[k]	12-bit	0.39–1.04	Up to 15
AATSR (Envisat-1)	ESA	2002–	500	1000	35 days[k]	12-bit	0.55–12	7
ASAR (Envisat-1)	ESA	2002–	100–400[i]	30–1000[i]	35 days[l]		Radar	N/A
HRC (CBERS-2B)	CBERS, China/Brazil	2007–	27	2.7	26 days	8-bit	0.45–0.85	1
CCD (CBERS-2B)	CBERS, China/Brazil	2007–	113	20	26 days	8-bit	0.45–0.73	5
TANSO-FTS (GOSAT/Ibuki)	JAXA, Japan	2009–	160[m]	10.5	3 days		0.758–14.3	4
LISS-4 (RESOURCESAT-2)	ISRO, India	2011–	70	5.8	24 days	10-bit	0.52–0.86	3
LISS-3 (RESOURCESAT-2)	ISRO, India	2011–	141	23.5	24 days	10-bit	0.52–1.70	4
AWiFS (RESOURCESAT-2)	ISRO, India	2011–	740	56	24 days	12-bit	0.52–1.70	4

(continued)

Table 2.1 (continued)

Sensor (Mission)	Organization[a]	Operation period	Swath width (km)	Spatial resolution (m)[b]	Temporal resolution	Radiometric resolution	Spectral resolution (μm)	Spectral bands
HRVIR (SPOT 4,5)	SPOT Image, France	1998–	60	10 (PAN), 20 (MS)	26 days[n]	8-bit	0.61–0.68 (PAN), 0.50–1.75	4 + PAN
VEGETATION (SPOT 4,5)	SPOT Image, France	1998–	2250	1150	26 days[n]	10-bit	0.43–1.75	4
HRG (SPOT 5)	SPOT Image, France	2002–	60	2.5–5 (PAN), 10 (VNIR), 20 (SWIR)	26 days[n]	8-bit	0.48–0.71 (PAN), 0.50–1.75	4 + PAN
IKONOS	GeoEye, USA	1999–	11.3	1 (PAN), 4 (MS)	3–5 days	11-bit	0.526–0.929 (PAN), 0.445–0.853	4 + PAN
Quickbird	DigitalGlobe, USA	2001–	18	0.65 (PAN), 2.62 (MS)	2.5–5.6 days	11-bit	0.405–1.053 (PAN), 0.43–0.918	4 + PAN
GeoEye-1	GeoEye, USA	2008–	15.2	0.41 (PAN), 1.65 (MS)	<3 days	11-bit	0.45–0.80 (PAN), 0.45–0.92	4 + PAN
WorldView-2	DigitalGlobe, USA	2009–	16.4	0.46 (PAN), 1.85 (MS)	1.1–3.7 days	11-bit	0.45–0.80 (PAN), 0.45–1.04	8 + PAN

[a] Organization acronyms: *NASA* National Aeronautics and Space Administration, *METI* Ministry of Economy, Trade, and Industry, *CNES* Centre National d'Études Spatiales, *NOAA* National Oceanic and Atmospheric Administration, *NESDIS* National Environmental Satellite, Data, and Information Service, *CSA* Canadian Space Agency, *ESA* European Space Agency, *CBERS* China-Brazil Earth Resources Satellite Program, *JAXA* Japanese Aerospace Exploration Agency, *ISRO* Indian Space Research Organization

[b] Acronyms used in describing sensor channels/configurations: *MS* multispectral, *TIR* Thermal infrared, *PAN* Panchromatic, *VNIR* Visible and near-infrared, *SWIR* Short-wave infrared

[c] The MSS sensor on Landsat 3 had a fifth spectral band for thermal infrared. The MSS sensors on other Landsat missions had four-band configurations

[d] Each receiver channel on the CALIOP sensor, a spaceborne LiDAR system, has dual 14-bit digitizers that jointly provide a 22-bit dynamic range

[e] The CALIOP sensor produces simultaneous laser pulses at two wavelengths, 0.532 and 1.064 μm

[f] The NOAA satellite program began in 1978. Currently, NOAA-19 is designated as the program's "operational" satellite, while NOAA-15, NOAA-16, NOAA-17, and NOAA-18 still transmit data as "standby" satellites

[g] While the AVHRR sensor has six spectral channels, only five are transmitted to the ground at any time. The bands designated 3A and 3B are transmitted only during daytime and nighttime, respectively

[h] Although the GOES satellite program has been active since 1975, only GOES-11 through GOES-15 are currently operational. GOES-11 was launched in 2000 but put into operation in 2006, replacing an earlier GOES satellite

[j] The radar systems in this table (i.e., RADARSAT-1, RADARSAT-2, and ASAR) operate in a variety of scan modes with different swath widths and spatial resolutions

[k] The orbit cycle of the Envisat-1 satellite is 35 days. The wide swaths of the MERIS and AATSR sensors permit more rapid revisits (approximately every three and every six days, respectively), but images are captured from different orbits and thus have different observation geometry, which may affect image processing

[l] The repeat cycle of the Envisat-1 satellite is 35 days, but more rapid revisits are possible with the ASAR depending on the latitude of the area of interest, requirements regarding the incidence angle (i.e., the angle between the radar beam and a line perpendicular to the ground), and desired scan mode

[m] The FTS sensor on the GOSAT satellite records 10.5 × 10.5 km images that, nominally, are spaced 150 km apart in a grid (see Kuze et al. 2009)

[n] The SPOT satellites have the capacity to record data off-nadir (i.e., to record data in areas that are not directly below the satellite). This may reduce revisit time to 2–3 days (1 day for the wide-swath VEGETATION sensor), but the images will have different observation geometry, which may affect image processing

Fig. 2.1 Landsat ETM+ (*left*) and Quickbird (*right*) images of the same area, demonstrating the large difference in spatial resolution between the two sensors (Landsat ETM+ multispectral resolution = 30 m; Quickbird multispectral resolution = 2.62 m). In the Quickbird image, individual buildings and minor road features are readily discernible

Table 2.2 Comparison of the spectral resolutions of the Landsat ETM+ and Quickbird sensors

Bandwidth (µm)

Spectral band	Landsat ETM+[a]	Quickbird
Panchromatic	0.52–0.9	0.405–1.053
Blue	0.45–0.515	0.43–0.545
Green	0.525–0.605	0.466–0.62
Red	0.63–0.69	0.59–0.71
Near-infrared	0.75–0.9	0.715–0.918
Mid-infrared	1.55–1.75	
Thermal infrared	10.4–12.5	
Mid-infrared	2.09–2.35	

[a] The ETM+ sensor is configured with two mid-infrared bands, designated bands 5 and 7, on either side of band 6, a thermal infrared band

as well as a panchromatic band. Table 2.2 compares the spectral bandwidths of the Landsat ETM+ and Quickbird sensors. Notably, in many geographic areas, ambient electromagnetic reflectance is strongly shaped by the distribution and condition of vegetation and water features, especially in the **visible to near-infrared** (**VNIR**) range (0.4–1.4 µm) of the electromagnetic spectrum to which many of the sensors in Table 2.1 are attuned.

Alternatively, NASA's Hyperion sensor, the first spaceborne hyperspectral sensor, provides a virtually continuous recording of electromagnetic reflectance across 220 narrow spectral bands in the VNIR and **short-wave infrared** (**SWIR**) wavelengths. Although the processing complexity of hyperspectral data has hindered its applicability in many land cover studies, these data provide a great deal of information about the unique reflectance behavior of vegetation stress and diversity. The high spectral resolution also provides an increased ability to discriminate between features that have similar response profiles, or **signatures**, across the electromagnetic spectrum, such as forest stands composed of different deciduous tree species (Chen and Hepner 2001).

Often, temporal resolution, or revisit time, is expressed in terms of days. For example, Landsat-7 has a 16-day orbit cycle, meaning that the satellite (and its ETM+ sensor) returns to a given location on Earth's surface every 16 days. However, some satellites, such as the Geostationary Operational Environmental Satellites (GOES), have a revisit cycle of less than an hour in certain locations. The GOES satellites, which are most commonly associated with meteorology and climate research, have a coarse, 8-km spatial resolution which gives them limited utility for local land cover studies. By contrast, the balance of spatial and temporal resolution achieved by NASA's Moderate Resolution Imaging Spectroradiometer (MODIS)—with a spatial resolution of 250–1000 m and a temporal resolution of 1–2 days—has made the sensor applicable to a variety of regional- to continental-scale research efforts.

In addition to the high spatial resolutions possible with commercial satellites such as WorldView-2, their onboard sensors also have high radiometric resolutions. Their 11-bit collection depth represents a substantial improvement over the 8-bit resolution typically exhibited by predecessors such as the Landsat sensors. Radiometric resolution refers to the **dynamic range** (i.e., the number of possible data values) recorded in each image band. Higher radiometric resolution means that a sensor is more sensitive to variations in electromagnetic reflectance (Fig. 2.2). This increased sensitivity has been shown to be very useful in the land cover classification of complex (i.e., urban and suburban) landscapes (Hester et al. 2008, 2010).

Ultimately, each of the four resolution types must be considered in light of the intended purpose of the data. For example, except near cities, most terrestrial landscapes are dominated either by vegetation, water, or other natural surfaces, even amid production land uses such as agriculture or mining. Spatial resolution might be relatively less important for mapping or monitoring efforts in these settings because of high landscape homogeneity. In heavily vegetated environments, data that are high in temporal and spectral resolution rather than spatial resolution, such as AV-HRR imagery, could be a powerful tool for evaluating seasonal or annual change in photosynthetic activity (Moulin et al. 1997; Stöckli and Vidale 2004). With respect to other types of analyses, hyperspectral data have demonstrated strong applicability for geological, mineralogical, and soil studies (e.g., Galvão et al. 2008; Weber et al. 2008). In addition, **radar** systems such as the Advanced Synthetic Aperture Radar (ASAR) on the Envisat-1 satellite can be useful for observing changes in sea surface conditions (Johannessen et al. 2008), while one of the major applications of Envisat-1's Medium Resolution Imaging Spectrometer (MERIS) sensor is for studying changes in ocean color (Cui et al. 2010). In short, there is no one "best" sensor for mapping a particular type of landscape. Although the sensors listed in Table 2.1 are only a sample of the instruments currently producing remotely sensed data, they illustrate the wide diversity of options that an analyst should consider before choosing a particular sensor.

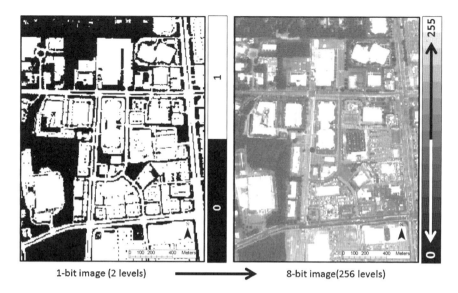

1-bit image (2 levels) ➡️ **8-bit image(256 levels)**

Fig. 2.2 The concept of radiometric resolution is illustrated by two images. The image on the left represents a 1-bit image, where two *brightness* values, or *grayscale* levels (i.e., *black* and *white*), are portrayed within the data. The image on the right represents an 8-bit image, where up to 256 *grayscale* levels are portrayed within the data. In contrast to these images, an 11-bit image holds up to 2,048 *grayscale* levels

Payloads and Platforms: An Overview

In remote sensing, the carrier of a sensor is known as its **platform**, while the sensor itself is the platform's **payload**. Remote sensing platforms are split into two categories: airborne and spaceborne. During the last two decades, there has been rapid proliferation of platforms in both categories. Here, we provide a brief overview of major platforms and sensor formats in use historically and today.

Airborne Platforms

Although the primary focus of this book is satellite-based remote sensing and the processing of associated digital data, until the emergence of satellite imagery in the 1970s, aerial photography served as the main data source when mapping phenomena on Earth's surface (Short 2010). Through time, "true" color and color-infrared film became economically feasible as alternatives to panchromatic film, and offered better capability for tasks such as vegetation classification (Caylor 2000). Until recently, digital cameras did not provide the same level of fine detail as film-based cameras, but many currently available digital aerial cameras have comparable spatial resolutions to film cameras, with similar formatting and scale (Morgan et al. 2010).

Fig. 2.3 A vertical aerial photograph (*left*) and an oblique aerial photograph (*right*). Images courtesy of Ohio Department of Transportation

Aerial photography may be either **vertical** or **oblique** (Fig. 2.3), depending on the orientation of the camera's optical axis relative to Earth. Vertical photos are captured with the camera's axis at an angle perpendicular or near-perpendicular ($90° \pm 3°$) to the ground. Oblique photos are typically captured with the camera's axis tilted more than $20°$ from vertical.

Because airborne platforms are not suited to capturing large geographic areas (e.g., $10,000$ km^2) at once, they have been replaced by spaceborne platforms for most broad-scale remote sensing projects (Short 2010). Nevertheless, aerial photography remains viable because it can be tailored to specific project needs (e.g., a particular spatial scale or resolution) in ways that satellite imagery cannot (Morgan et al. 2010). Aerial photography continues to be the foundation of many national-scale mapping efforts, such as the National Agricultural Imagery Program (NAIP), which has the goal of providing regularly updated, geo-referenced imagery to the public (USDA FSA 2010). In addition, the development of image processing techniques for satellite imagery has in turn expanded the range of automated techniques that may be applied to digital aerial photos (Morgan et al. 2010), including **object-oriented classification** (see Chap. 3). Various non-photographic sensors have also been implemented via airborne platforms, including **active remote sensing systems** such as radar and **LiDAR**. Radar (short for RAdio Detection and Ranging) systems operate within the microwave portion of the electromagnetic spectrum (Lillesand et al. 2008). Essentially, a radar sensor works by emitting radio waves from an antenna that bounce off target features on Earth, and then the sensor records the returned energy. The target features may be distinguished from one another by their differing effects on the returned signals. Radar sensors can capture images day or night, and in all weather conditions (e.g., through cloud cover or smoke). For this reason, radar imaging is often used in disaster management, for instance in detecting ocean oil spills (Jha et al. 2008; Klemas 2010). Radar **interferometry** involves the simultaneous use of two antennae to generate radio waves and collect returned signals; incorporating the distance between these antennae during data processing facilitates topographic

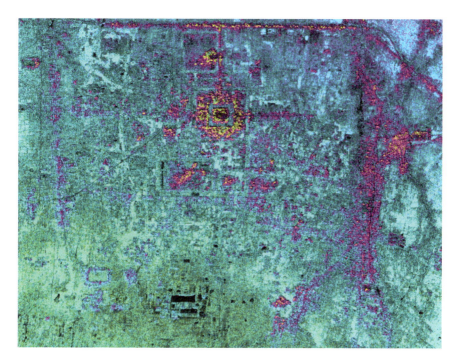

Fig. 2.4 Topographic image of Hariharalaya, the nineth century capital of the Khmer Empire in Cambodia. The image was captured by the NASA Airborne Synthetic Aperture Radar (AIRSAR) sensor, operating in interferometric mode. Colors represent elevation contours. Image courtesy of NASA Jet Propulsion Laboratory

mapping (Massonnet and Fiegl 1998). For instance, Fig. 2.4 shows an application of airborne radar interferometry to develop a **digital elevation model**, or **DEM**, of the area surrounding an ancient city site in Cambodia.

LiDAR (short for Light Detection And Ranging) systems resemble radar in that the output images are generated based on the amount of sensor-emitted energy that is returned from features on the ground. However, unlike radar, LiDAR sensors emit laser pulses (typically in the near-infrared portion of the electromagnetic spectrum) at a very high rate (i.e., 10,000–100,000 pulses per second) (Reutebuch et al. 2005). In turn, the three-dimensional positions of targeted objects are determined based on the time it takes for pulses to return the sensor. The typical LiDAR output image is a topographic data set with very accurate vertical measurements ($\pm$ 10–15 cm error) (Charlton et al. 2003; Reutebuch et al. 2005). Most airborne LiDAR systems can record multiple returns (Fig. 2.5a) from the same laser pulse in cases where an object does not completely block the pulse's path, allowing it to continue to another object closer to the ground (Reutebuch et al. 2005). This feature makes LiDAR systems useful for applications such as the characterization of forest canopy structure (Fig. 2.5b) (Reutebuch et al. 2005; Lim et al. 2003). However, LiDAR systems are sensitive to weather and other conditions that interfere with the laser pulses.

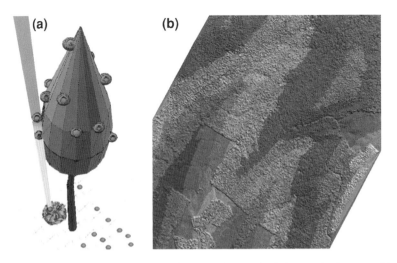

Fig. 2.5 **a** Conceptual rendering of multiple returns from a single laser pulse from an airborne LiDAR system. **b** Forest canopy surface recorded by LiDAR; colors indicate elevation contours. Image credit: Mississippi State University, College of Forest Resources, Measurements & Spatial Technologies Laboratory

Fig. 2.6 A hyperspectral image data "cube" of Moffit Field, California, captured by the AVIRIS sensor. Moffit Field is at the southern end of San Francisco Bay. The top of the image cube is a false color composite showing evaporation ponds (*center* and *lower right*) as well the Moffit Field airport (*left*). The sides of the cubes show the edges of each image corresponding to one of the sensor's 224 spectral bands. Image courtesy of NASA Jet Propulsion Laboratory

For the past few decades, multispectral scanners resembling the passive sensors found on many Earth-observing satellites have been mounted on airborne platforms. More recently, there has also been increased application of airborne hyperspectral sensors. For example, the AVIRIS (Airborne Visible Infrared Imaging Spectrometer) instrument (Fig. 2.6), operated by NASA, collects data from 224 contiguous bands spanning the ultraviolet to near infrared; the spectral resolution of each AVIRIS band is approximately 0.01 µm. This high degree of spectral resolution enables analysts to make subtle distinctions regarding objects or areas of interest that are not possible with multispectral data. For instance, hyperspectral imagery has been used to detect early stages of forest damage and decline caused by air pollution (Campbell et al. 2004). However, because of the very large data volume associated with the typical hyperspectral image, an analyst must often use statistical techniques (e.g., principal components analysis) to reduce the dimensionality prior to further image processing (Harsanyi and Chang 1994).

Spaceborne Platforms

An exhaustive list of all Earth-observing satellites that are currently in operation is beyond the scope of this book. Instead, we provide basic details about some prominent government and commercial satellites from the U.S. and elsewhere (mostly listed in Table 2.1). We also highlight promising developments in spaceborne remote sensing (e.g., the availability of satellite-based hyperspectral and LiDAR sensors).

NASA Satellites and Satellite Programs

Essentially, satellite remote sensing began with NASA's Landsat Program. Landsat-1, launched in 1972, was the first in a series of satellites associated with the long-running program, the latest being Landsat-7, launched in April 1999 (Short 2010; Jensen 2005). Besides, Landsat-7, only Landsat-5 (launched in March 1984) remains operational. Both Landsat-5 and Landsat-7 have a 16-day revisit time.

Landsat-5's primary on-board sensor, the Thematic Mapper (TM), was the predecessor of the Enhanced Thematic Mapper (ETM+) on Landsat-7. The ETM+ sensor incorporates two major improvements on the TM sensor: a higher-resolution thermal infrared band (60 m versus 120 m for the TM sensor) as well as the novel inclusion of a 15-m panchromatic band. The ETM+ sensor requires roughly 11,000 scenes to completely image Earth, excluding the polar regions (Short 2010).

With its various sensors, the Landsat Program has provided the longest—by far—continuous and comprehensive record of Earth imagery. Each spectral band of the ETM+ sensor has utility for certain environmental applications (Table 2.3). Figure 2.7 is an ETM+ image of the Ganges River Delta; this false-color composite was made using green, infrared, and blue wavelengths. In addition, the entire Landsat image archive has been made available to the public via the Internet, facilitating long-term change analyses.

Landsat-7 is considered part of NASA's Earth Observing System (EOS) mission, which also includes the paired satellites Terra (launched December 1999) and Aqua (launched May 2002). Together, the Aqua and Terra satellites cover the mid to higher latitudes of Earth four times daily. Both satellites carry the Moderate Resolution Imaging Spectroradiometer (MODIS) sensor. The MODIS sensor

Table 2.3 Capabilities and applications of ETM+ spectral bands (adapted from Jensen 2005)

Band	Capabilities/Applications
Band 1 (blue)	Penetrating water bodies; analysis of land-use, soil, and vegetation
Band 2 (green)	Green reflectance of healthy vegetation
Band 3 (red)	Vegetation discrimination; delineation of soil and geologic boundaries
Band 4 (near-infrared)	Crop identification; emphasizes soil–crop and land–water contrasts
Band 5 (mid-infrared)	Drought studies; discrimination between clouds, snow, and ice
Band 6 (thermal infrared)	Locating geothermal activity; vegetation stress analysis; soil moisture studies; detection of urban heat islands
Band 7 (mid-infrared)	Discrimination of geologic rock formations

records data in 36 spectral bands spanning the visible to thermal infrared, with a unique configuration: bands 1 and 2 have spatial resolutions of 250 m, bands 3–7 have spatial resolutions of 500 m, and bands 8–36 have spatial resolutions of 1 km. MODIS data have wide applicability in terrestrial, atmospheric, and marine contexts, primarily for broad-scale (i.e., continental-scale) analyses; some of these applications are highlighted in Chaps. 4, 5 and 6.

The Advanced Spaceborne Thermal Emission and Reflection Radiometer (ASTER) sensor aboard EOS Terra is actually a combination of three radiometers that yield simultaneous and co-registered image data of differing spatial resolutions: three VNIR bands with a spatial resolution of 15 m; six SWIR bands with a spatial resolution of 30 m; and five TIR bands with a spatial resolution of 90 m. While ASTER imagery has regularly been used to study land surface temperature and thermal phenomena such as surface emissivity (e.g., Schmugge et al. 2002), it has also been applied for urban land cover classification (Chen et al. 2007).

NASA launched Earth Observing-1 (EO-1) satellite into orbit in November 2000. Of its various instruments, the Hyperion imaging spectrometer, with its 220 spectral bands, is of particular note. Hyperion data have been used for applications such as analysis of desertification (Asner and Heidebrecht 2003) and the detection of fungal disease in sugarcane (Apan et al. 2004). Figure 2.8 is a Hyperion image, captured in October 2007, of a large wildfire complex in northern San Diego County, California; using three short-wave infrared channels, the sensor was able to cut through dense smoke plumes to show the locations of actively burning fires. However, Hyperion data tend to have a higher signal-to-noise ratio relative to data from airborne hyperspectral sensors such as AVIRIS, which limits their utility for fine-scale mapping and classification (Apan 2004; Kruse et al. 2003). Future spaceborne hyperspectral sensors will likely include technological improvements to address this signal-to-noise ratio limitation.

In May 2003, the ETM+ sensor aboard Landsat-7 experienced a failure of its scan-line corrector (SLC) mechanism. This uncorrectable failure results in data gaps on the left and right sides of each recorded image (Williams et al. 2006). These data gaps are filled using data from another ETM+ image captured close in

Fig. 2.7 Landsat ETM+ image of the Ganges River Delta, captured in February 2000. The delta is dominated by swamp forest. Image courtesy of U.S. Geological Survey, EROS Data Center, Satellite Systems Branch

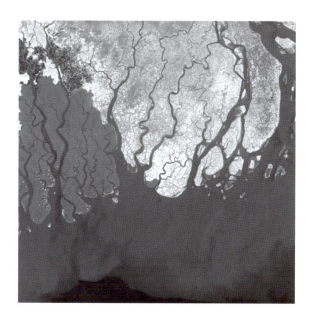

Fig. 2.8 Hyperion image of the Witch Wildfire in northern San Diego County, California, captured in October 2007. Image courtesy of NASA EO-1 Team

time, but the SLC problem—combined with the fact that the only other operational Landsat satellite, Landsat-5, is approaching 30 years of service—emphasizes that the Landsat mission is likely approaching its end. Another sensor on board the EO-1 satellite, the Advanced Land Imager (ALI), was developed as a prototype for sensors that could be carried on future Landsat missions (Chander et al. 2004).

The CALIPSO satellite is jointly operated by NASA and Centre National d'Études Spatiales (CNES), the French government space agency. A key instrument on CALIPSO is the Cloud-Aerosol LiDAR with Orthogonal Polarization (CALIOP) sensor, which represents one of very few operational space-borne

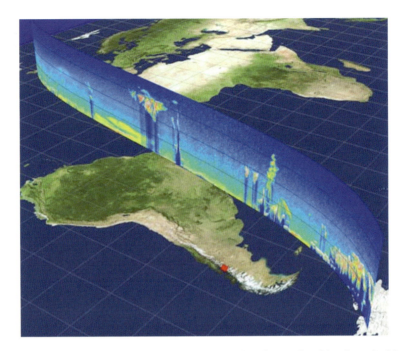

Fig. 2.9 CALIOP image showing vertical profile of ash and aerosols arising from the May 6, 2008 eruption of Chile's Chaitén volcano (*red square*). The top of the aerosol column is approximately 16 km altitude. Image courtesy K. Severance and C. Trepte, NASA Langley Research Center

LiDAR systems. As its name suggests, the CALIOP sensor is used for analysis of a variety of atmospheric phenomena, including dust storms (Liu et al. 2008). Figure 2.9 is a CALIOP image captured May 7, 2008, the day after an eruption of Chile's Chaitén Volcano. Among the phenomena detected by CALIOP was the presence of aerosols in the stratosphere, drifting over southeastern Australia, which suggests rapid, long-distance transport of fine ash (Carn et al. 2009).

Other Government Satellites and Satellite Programs

The **Advanced Very High Resolution Radiometer (AVHRR)** is a sensor mounted on a series of satellites operated by the U.S. National Oceanic and Atmospheric Administration (NOAA). The latest in the series, NOAA-19, was launched in 2009. NOAA-19 carries the third version of the AVHRR sensor, with six spectral bands. Only five bands are transmitted to the ground at any given time; the bands designated 3A and 3B are transmitted during daytime and nighttime, respectively. The AVHRR sensor was primarily developed for oceanographic and meteorological applications, but it has also been used for land cover mapping at broad scales (e.g., Loveland et al. 2000), especially because of its twice-a-day temporal coverage (Jensen 2005).

The NOAA satellites have **polar orbits**, meaning that they pass over the North and South Poles during each trip around Earth. Many other satellites, such as Landsat-7, have **near-polar orbits**. Polar- and near-polar-orbiting satellites are typically **sun-synchronous**, which means that the satellite passes over a given location on Earth at approximately the same local time each day. This characteristic enables regular data collection at consistent times as well as long-term comparisons.

Alternatively, some satellites are placed into a **geostationary orbit**, where the satellite remains fixed in a particular location above Earth (commonly the Equator), and orbits in the direction (and at the speed) of the planet's rotation. One advantage of geostationary satellites is that it easy for ground-based antennae to communicate with them. For this reason, this is the preferred orbit for many weather monitoring satellites, including the Geostationary Operational Environmental Satellite (GOES) system, which serves as the primary information source for the U.S. National Weather Service. The sensor (i.e., the I-M Imager instrument) aboard each GOES satellite records individual images covering approximately 25% of Earth's surface (Jensen 2005). Currently, five GOES satellites are active: GOES-11, designated "GOES West", which is located at 135° W longitude; GOES-12, located at 60° W and currently tasked with imaging South America; GOES-13, designated "GOES East", which is situated at 45° W; GOES-14, which is in on-orbit storage status; and GOES-15, which is on stand-by (NOAA 2011).

The Canadian Space Agency (CSA) launched its first Earth-observing satellite, RADARSAT-1, in November 1995, followed by the launch of RADARSAT-2 in December 2007. The primary payload on each of these satellites is a **synthetic aperture radar (SAR)** imaging system. A SAR system utilizes the motion of the platform that carries it (e.g., a satellite) to simulate a large antenna, or aperture. In short, as the platform moves, the system transmits multiple, successive radio pulses toward targeted objects or areas. The returned signals are then synthesized into a single image with higher spatial resolution than could be captured with a smaller, physical antenna.

The RADARSAT satellites have the all-weather capabilities of airborne radar systems, but with the added advantage of frequent revisits (every 24 days) of targeted areas. (RADARSAT-2 has side-looking modes that can further reduce revisit time.) The SAR systems on the RADARSAT satellites also have multiple beam modes that essentially trade swath size for improved spatial resolution. A particular strength of satellite-based SAR imagery is the ability to measure ocean waves and currents over large areas (Goldstein et al. 1989); for instance, Fig. 2.10 is a RADARSAT image showing wave patterns in the Pacific Ocean around Point Reyes, California.

The European Space Agency (ESA) launched its first Earth-observing satellite, ERS-1, in 1991, followed by ERS-2 in 1995 and Envisat-1 in 2002 (the latter two are still operational.) Envisat-1 has nine on-board instruments, including the Medium Resolution Imaging Spectrometer (MERIS). As noted earlier, one of the main applications of the MERIS sensor has been for analyzing ocean color

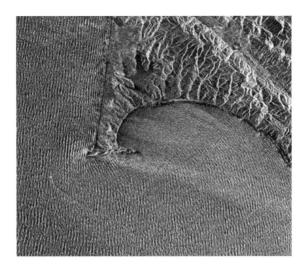

Fig. 2.10 RADARSAT image recorded near Point Reyes, California. Image courtesy of NASA

Fig. 2.11 MERIS image, captured in August 2003, of the coastline of Bangladesh. Image courtesy of the European Space Agency

changes as an indicator of water quality (Cui et al. 2010). For example, Fig. 2.11 is a MERIS image of the coastline of Bangladesh, showing extensive sediments flowing into the Bay of Bengal.

Several nations outside North America and Europe have also established successful satellite programs. For example, the Indian Space Research Organization (ISRO) has constructed and launched 18 Earth-observing satellites, starting with Indian Resources Satellite (IRS) 1A in 1988. Its latest satellite, Resourcesat-2, carries three multispectral sensors with differing spatial resolutions: the 5.8-m, three-spectral-band Linear Imaging Self Scanner (LISS)-4; the 23.4-m, 4-band Linear Imaging Self Scanner (LISS)-3; and the 56-m, four-band Advanced Wide

Field Sensor (AWiFS). Together, these three sensors offer a high degree of analytical flexibility (Seshadri et al. 2005), making the satellite's data potentially suitable for a variety of tasks, as demonstrated by the use of Resourcesat imagery for studying coastal dynamics and water quality (Rajawat et al. 2007) and monitoring snow cover (Kulkarni et al. 2006).

The China-Brazil Earth Resources Satellite (CBERS) program was established by a cooperative agreement between the two countries in 1988. The CBERS-2B satellite, launched in 2007, carries three instruments, including a medium-resolution (20-m), multispectral charge coupled device (CCD) camera, which has been suggested as a low-cost supplement or alternative to other medium-resolution sensors such as Landsat ETM+ or Resourcesat-2's LISS-3 (Ponzoni and Albuquerque 2008). For example, Redo (2012) described a study of deforestation in Bolivia over a 33-year period (1975–2008), where data from multiple Landsat missions (i.e., MSS, TM, and ETM+ imagery) were combined with data from CBERS-2 and CBERS-2B to analyze a $63,000$-km^2 area known as the Corredor Bioceánico.

The Japanese Aerospace Exploration Agency (JAXA) was formed in 2003 from a merger of three organizations: the National Space Development Agency of Japan (NASDA), the National Aerospace Laboratory (NAL), and the Institute of Space and Aeronautical Science (ISAS). One of JAXA's ongoing missions is the Greenhouse Gases Observing Satellite (GOSAT), or "Ibuki". Launched in 2009, GOSAT is the first satellite designed to remotely measure atmospheric concentrations of carbon dioxide (CO_2) and methane (CH_4) near Earth's surface (Morino et al. 2010; Kuze et al. 2009).

Commercial Satellites

Spot Image is a public company started by the French Space Agency (CNES) in 1982 (and now a subsidiary of EADS Astrium). Spot Image manages the SPOT Earth-observing satellites. The first satellite in the series, SPOT-1, was launched in 1986; SPOT-4 (launched in 1998) and SPOT-5 (launched in 2002) are currently in operation. Both SPOT-4 and SPOT-5 carry a "high-resolution" VNIR (HRVIR) sensor similar in capability to Landsat ETM+ (see Table 2.1), as well as a coarse-resolution VEGETATION sensor intended for regional mapping studies (Stibig et al. 2007). Unlike the Landsat-mounted sensors, the SPOT sensors are configured to permit off-nadir viewing (Jensen 2005), which can reduce revisit time from 26 days (i.e., the satellites' orbit cycle) to 2–3 days for the HRVIR sensor and one day for the VEGETATION sensor.

The SPOT-5 satellite also carries two High-Resolution Geometrical (HRG) instruments. Images from these 5-m panchromatic sensors have proven sufficient for fine-scale analyses such as the automatic detection of ships for monitoring of fisheries (Corbane et al. 2008). Furthermore, the two HRG sensors can be operated together in "super mode" to generate 2.5-m panchromatic imagery (Pasqualini et al. 2005).

The success of Spot Image laid the groundwork for the other commercially operated satellites that, starting with IKONOS in 1999, have provided panchromatic and multispectral imagery with very high spatial resolutions. (Indeed, in keeping with the trend, Spot Image's forthcoming Pléiades satellite will provide 0.5-m panchromatic and 2-m multispectral data.) Through a series of mergers and acquisitions, two U.S. companies have become the main contemporary providers of high-spatial-resolution imagery: GeoEye, which manages the IKONOS and GeoEye-1 satellites; and DigitalGlobe, which operates the Quickbird and World-View-2 satellites (see Table 2.1 for sensor specifications). As noted earlier in this chapter, there are clearly many potential applications for these data. Nevertheless, it is important to acknowledge that such images are not well suited to standard classification techniques (e.g., maximum likelihood classification), primarily because objects and areas of interest are often represented by multiple pixels with fairly high variability in their spectral values (e.g., due to shadows) (Blaschke 2010; Yu et al. 2006). Instead, object-oriented classification techniques (see Chap. 3) are increasingly popular for application to high-spatial-resolution images. Another issue for commercial satellite imagery is the cost to the user, especially given the free availability of other satellite imagery (e.g., Landsat) that, while of coarser spatial resolution, may be sufficient for analysis. In any case, users of remotely sensed data should carefully consider the objectives of their particular projects when selecting the imagery most likely to meet those objectives.

References

A. Apan, A. Held, S. Phinn, J. Markley, Detecting sugarcane 'orange rust' disease using EO-1 Hyperion hyperspectral imagery. Int. J. Remote Sens. **25**, 489–498 (2004)

G.P. Asner, K.B. Heidebrecht, Imaging spectroscopy for desertification studies: comparing AVIRIS and EO-1 Hyperion in Argentina drylands. IEEE Trans. Geosci. Remote. Sens. **41**, 1283–1296 (2003)

T. Blaschke, Object based image analysis for remote sensing. J. Photogramm. Remote Sens. **65**, 2–16 (2010)

P.K.E. Campbell, B.N. Rock, M.E. Martin, C.D. Neefus, J.R. Irons, E.M. Middleton, J. Albrechtova, Detection of initial damage in Norway spruce canopies using hyperspectral airborne data. Int. J. Remote Sens. **25**, 5557–5583 (2004)

S.A. Carn, J.S. Pallister, L. Lara, J.W. Ewert, S. Watt, A.J. Prata, R.J. Thomas, G. Villarosa, The unexpected awakening of Chaitén volcano, Chile Eos **90**, 205–206 (2009)

J. Caylor, Aerial photography in the next decade. J. For. **98**(6), 17–19 (2000)

G. Chander, D.J. Meyer, D.L. Helder, Cross calibration of the Landsat-7 ETM+ and EO-1 ALI sensor. IEEE Trans. Geosci. Remote Sens. **42**, 2821–2831 (2004)

M.E. Charlton, A.R.G. Large, I.C. Fuller, Application of airborne LiDAR in river environments: the river Coquet, Northumberland, UK. Earth Surf. Proc. Land. **28**, 299–306 (2003)

J. Chen, G.F. Hepner, Investigation of imaging spectroscopy for discriminating urban land covers and surface materials, in *Proceedings of the 2001 AVIRIS Earth Science and Applications Workshop*, Palo Alto, CA. NASA JPL Publication 02–1. 2001. http://popo.jpl.nasa.gov/docs/work-shops/01_docs/2001Chen_web.pdf

Y. Chen, P. Shi, T. Fung, J. Wang, X. Li, Object-oriented classification for urban land cover mapping with ASTER imagery. Int. J. Remote Sens. **28**, 4645–4651 (2007)

C. Corbane, F. Marre, M. Petit, Using SPOT-5 HRG data in panchromatic mode for operational detection of small ships in tropical area. Sensors **8**, 2959–2973 (2008)

T. Cui, J. Zhang, S. Groom, L. Sun, T. Smyth, S. Sathyendranath, Validation of MERIS ocean-color products in the Bohai Sea; A case study for turbid coastal waters. Remote Sens. Environ. **114**, 2326–2336 (2010)

L.S. Galvão, A.R. Formaggio, E.G. Couto, D.A. Roberts, Relationships between the mineralogical and chemical composition of tropical soils and topography from hyperspectral remote sensing data. ISPRS J. Photogramm. Remote Sens. **63**, 259–271 (2008)

R.M. Goldstein, T.P. Barnett, H.A. Zebker, Remote sensing of ocean currents. Science **246**, 1282–1285 (1989)

J.C. Harsanyi, C.-I. Chang, Hyperspectral image classification and dimensionality reduction: an orthogonal subspace projection approach. IEEE Trans. Geosci. Remote Sens. **32**, 779–785 (1994)

D.B. Hester, H.I. Cakir, S.A.C. Nelson, S. Khorram, Per-pixel classification of high spatial resolution satellite imagery for urban land-cover mapping. Photogramm. Eng. Remote Sens. **74**, 463–471 (2008)

D.B. Hester, S.A.C. Nelson, H.I. Cakir, S. Khorram, H. Cheshire, High-resolution land cover change detection based on fuzzy uncertainty analysis and change reasoning. Int. J. Remote Sens. **31**, 455–475 (2010)

J.R. Jensen, *Introductory Digital Image Processing: A Remote Sensing Perspective.*, 3rd edn. (Prentice Hall, Upper Saddle River, 2005), p. 526

M.N. Jha, J. Levy, Y. Gao, Advances in remote sensing for oil spill disaster management: state-of-the-art sensors technology for oil spill surveillance. Sensors **8**, 236–255 (2008)

J.A. Johannessen, B. Chapron, F. Collard, V. Kudryavtsev, A. Mouche, D. Akimov, K.-F. Dagestad, Direct ocean surface velocity measurements from space: Improved quantitative interpretation of Envisat ASAR observations. Geophys. Res. Lett. **35**, L22608 (2008)

V. Klemas, Tracking oil slicks and predicting their trajectories using remote sensors and models: case studies of the Sea Princess and Deepwater Horizon oil spills. J. Coastal Res. **26**, 789–797 (2010)

F.A. Kruse, J.W. Boardman, J.F. Huntington, Comparison of airborne hyperspectral data and EO-1 Hyperion for mineral mapping. IEEE Trans. Geosci. Remote Sen. **41**, 1388–1400 (2003)

A.V. Kulkarni, S.K. Singh, P. Mathur, V.D. Mishra, Algorithm to monitor snow cover using AWiFS data of Resourcesat-1 for the Himalayan region. Int. J. Remote Sens. **27**, 2449–2457 (2006)

A. Kuze, H. Suto, M. Nakajima, T. Hamazaki, Thermal and near infrared sensor for carbon observation Fourier-transform spectrometer on the Greenhouse Gases Observing Satellite for greenhouse gases monitoring. Appl. Opt. **48**, 6716–6733 (2009)

T. Lillesand, R. Kiefer, J. Chipman, *Remote Sensing and Image Interpretation*, 6th edn. (Wiley, New York, 2008), p. 763

K. Lim, P. Treitz, M. Wulder, B. St-Onge, M. Flood, LiDAR remote sensing of forest structure. Prog. Phys. Geogr. **27**, 88–106 (2003)

Z. Liu, A. Omar, M. Vaughan, J. Hair, C. Kittaka, Y. Hu, K. Powell, C. Trepte, D. Winker, C. Hostetler, R. Ferrare, R. Pierce, CALIPSO lidar observations of the optical properties of Saharan Dust: a case study of long-range transport. J. Geophys. Res. **113**, D07207 (2008)

T.R. Loveland, B.C. Reed, J.F. Brown, D.O. Ohlen, Z. Zhu, L. Yang, J.W. Merchant, Development of a global land cover characteristics database and IGBP DISCover from 1 km AVHRR data. Int. J. Remote Sens. **21**, 1303–1330 (2000)

D. Massonnet, K.L. Feigl, Radar interferometry and its application to changes in Earth's surface. Rev. Geophys. **36**, 441–500 (1998)

J.L. Morgan, S.E. Gergel, N.C. Coops, Aerial photography: a rapidly evolving tool for ecological management. Bioscience **60**, 47–59 (2010)

I. Morino, O. Uchino, M. Inoue, Y. Yoshida, T. Yokota, P.O. Wennberg, G.C. Toon, D. Wunch, C.M. Roehl, J. Notholt, T. Warneke, J. Messerschmidt, D.W.T. Griffith, N.M. Deutscher, V. Sherlock, B. Connor, J. Robinson, R. Sussman, M. Rettinger, Preliminary validation of column-averaged volume mixing ratios of carbon dioxide and methane retrieved from GOSAT short-wavelength infrared spectra. Atmospheric. Meas. Tech. Discuss. **3**, 5613–5643 (2010)

S. Moulin, L. Kergoat, N. Viovy, G. Dedieu, Global-scale assessment of vegetation phenology using NOAA/AVHRR satellite measurements. J. Clim. **10**, 1154–1170 (1997)

National Oceanic and Atmospheric Administration (NOAA), GOES Spacecraft Status Main Page [web page]. NOAA National Environmental Satellite, Data, and Information Service (NESDIS), Office of Satellite Operations (2011), http://www.oso.noaa.gov/goesstatus/

V. Pasqualini, C. Pergent-Martini, G. Pergent, M. Agreil, G. Skoufas, L. Sourbes, A. Tsirika, Use of SPOT 5 for mapping seagrasses: An application to *Posidonia oceanica*. Remote Sens. Environ. **94**, 39–45 (2005)

F.J. Ponzoni, B.F.C. Albuquerque, Pre-launch absolute calibration of CCD/CBERS-2B sensor. Sensors **8**, 6557–6565 (2008)

A.S. Rajawat, M. Gupta, B.C. Acharya, S. Nayak, Impact of new mouth opening on morphology and water quality of the Chilika Lagoon—a study based on Resourcesat-1 LISS-III and AWiFS and IRS-1D LISS-III data. Int. J. Remote Sens. **28**, 905–923 (2007)

D. Redo, Mapping land-use and land-cover change along Bolivia's Corredor Bioceánico with CBERS and the Landsat series: 1975–2008. Int. J. Remote Sens. 33, 1881–1904 (2012)

S.E. Reutebuch, H.-E. Andersen, R.J. McGaughey, Light detection and ranging (LIDAR): an emerging tool for multiple resource inventory. J. For. **103**, 286–292 (2005)

J. Rogan, D.M. Chen, Remote sensing technology for mapping and monitoring land-cover and land-use change. Prog. Plan. **61**, 301–325 (2004)

T. Schmugge, A. French, J.C. Ritchie, A. Rango, H. Pelgrum, Temperature and emissivity separation from multispectral thermal infrared observations. Remote Sens. Environ. **79**, 189–198 (2002)

K.S.V. Seshadri, M. Rao, V. Jayaraman, K. Thyagarajan, K.R. Sridhara Murthi, Acta Astronautica **57**, 534–539 (2005)

N.M. Short, The remote sensing tutorial [web site]. National Aeronautics and Space Administration (NASA), Goddard Space Flight Center (2010), http://rst.gsfc.nasa.gov/

H.-J. Stibig, A.S. Belward, P.S. Roy, U. Rosalina-Wasrin, S. Agrawal, P.K. Joshi, Hildanus, R. Beuchle, S. Fritz, S. Mubareka, C. Giri, A land-cover map for South and Southeast Asia derived from SPOT-VEGETATION data. J. Biogeogr 34: 625–637 (2007)

R. Stöckli, P.L. Vidale, European plant phenology and climate as seen in a 20-year AVHRR land-surface parameter dataset. Int. J. Remote Sens. **25**, 3303–3330 (2004)

USDA Farm Service Agency (FSA), Imagery programs—NAIP imagery [web site]. U.S. Department of Agriculture, Farm Service Agency, Aerial Photography Field Office (2010), http://www.fsa.usda.gov/FSA/apfoapp?area=home&subject=prog&topic=nai

B. Weber, C. Olehowski, T. Knerr, J. Hill, K. Deutschewitz, D.C.J. Wessels, B. Eitel, B. Büdel, A new approach for mapping of Biological Soil Crusts in semidesert areas with hyperspectral imagery. Remote Sens. Environ. **112**, 2187–2201 (2008)

D.L. Williams, S. Goward, T. Arvidson, Landsat: yesterday, today, and tomorrow. Photogramm. Eng. Remote Sens. **72**, 1171–1178 (2006)

Q. Yu, P. Gong, N. Clinton, G. Biging, M. Kelly, D. Schirokauer, Object-based detailed vegetation classification with airborne high spatial resolution remote sensing imagery. Photogramm. Eng. Remote Sens. **72**, 799–811 (2006)

Relevant websites

NASA's Visible Earth, highlighting numerous application examples of satellite-based sensors: http://visibleearth.nasa.gov

Landsat image archive: http://landsat.usgs.gov/

Canadian Space Agency (CSA): http://www.asc-csa.gc.ca

European Space Agency (ESA): http://www.esa.int
Indian Space Research Organization (ISRO): http://www.isro.org
China-Brazil Earth Resources Satellite (CBERS) Program: http://www.cbers.inpe.br/en/
    index_en.htm
Japan Aerospace Exploration Agency (JAXA): http://www.jaxa.jp/index_e.html
EADS Astrium (Spot Image): http://www.astrium-geo.com
SPOT VEGETATION Free Distribution Site: http://free.vgt.vito.be/
DigitalGlobe: http://www.digitalglobe.com
GeoEye: http://www.geoeye.com

Chapter 3
Data Processing Tools

In this chapter, we discuss the tools and techniques that are conventionally used to process remotely sensed data into geospatial outputs (e.g., classified maps) for further analysis and application. This chapter focuses on the processing of moderate-spatial-resolution, multispectral digital image data, such as the data captured by the Landsat ETM+ sensor. We do not cover the processing of other categories of remotely sensed data such as radar and hyperspectral data; each requires specialized processing approaches that are beyond the scope of this book. Nevertheless, readers interested in learning more about the processing of these other data types should consult additional reading as referenced.

Briefly, the processing of multispectral image data is typically divided into three stages: preprocessing, processing, and post-processing. Each of these stages has its own distinct set of common tools and techniques. However, before discussing any of the stages of data processing, it is important to first understand the more fundamental task of image display.

Display of Multispectral Image Data

Raw image data are most commonly displayed by image processing software in the form of individual bands, or instead, as "true" or "false" color composites. Any individual band of a multispectral digital image can be displayed as grayscale, where the lowest-value pixels are displayed as black, the highest-value pixels are displayed as white, and pixels with intermediate values are displayed in corresponding shades of gray. Alternatively, true and false color composites will display no more than three image bands at a time, each matched to one of three primary colors: blue, green, and red. As perhaps expected, a true color composite (TCC) image displays the blue band from a raw multispectral image using the blue color ramp, the green band from the image using the green color ramp, and the red band using the red color ramp. Essentially, a TCC depicts its features in natural color. In contrast, a false color composite (FCC) displays the combination of any three bands from a multispectral

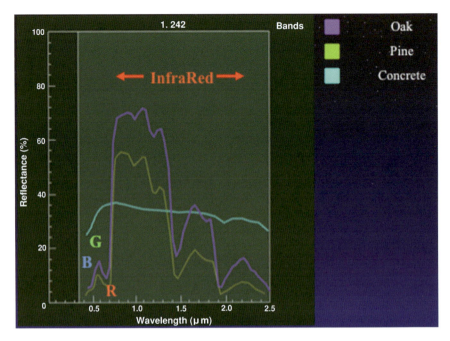

Fig. 3.1 Reflectance properties (i.e., spectral response curves) of oak forest, pine forest, and concrete, for wavelengths ranging from the visible to infrared portions of the electromagnetic spectrum

image other than the true-color combination. In the standard FCC, the green band of the input image is displayed using the blue color ramp, the red image band is displayed using the green color ramp, and a near-infrared band from the input image (e.g., Band 4 from a Landsat TM image) is displayed using the red color ramp (Khorram et al. in press). The FCC combination is popular because it highlights the presence of healthy green vegetation. Figure 3.1 displays line graphs of the reflectance properties (also called spectral response curves) of oak forest, pine forest, and concrete over a range of wavelengths from the visible to infrared portions of the electromagnetic spectrum. You will note that the two forest vegetation classes have much higher reflectance in near-infrared wavelengths than concrete. Thus, when an image's near-infrared band is displayed using a red color ramp in a FCC, live vegetation will appear red in the composite image. Figure 3.2 provides a side-by-side comparison of a TCC and FCC, constructed from Quickbird image data, for a study area in North Carolina.

Preprocessing Image Data

A remote sensing analyst employs preprocessing operations in order to prepare the best possible input data for the actual image processing stage. Basically, the

Fig. 3.2 An example TCC (*left*) and standard FCC (*right*) of a study area in coastal North Carolina based on Quickbird satellite data. Note that vegetation appears in various shades of green in the TCC and in various shades of red in the standard FCC

operations serve one of two purposes: (1) to minimize distortions and/or errors in an image that could hinder successful classification; or (2) to extract or enhance an image's most critical information, thus making classification more straightforward.

Geometric Correction

Remotely sensed images may contain two types of geometric distortions: systematic and non-systematic. Systematic distortions are due to image motion caused by forward movement of the aircraft or spacecraft, variations in mirror scanning rate, panoramic distortions, variations in platform velocity, and distortions due to the curvature of Earth. Non-systematic distortions are due to sensor malfunctions or variations in satellite altitude and attitude. Many systematic errors are already removed in commercially available data. The most common techniques for removing the remaining systematic and nonsystematic distortions are image-to-map rectification and image-to-image registration (i.e., geographic matching to existing spatial data), which involves the selection of a large number of well-defined ground control points shared by the target image and the reference image or map.

Atmospheric Correction

Suspended particles or other materials in the atmosphere at the time of data acquisition may change the data (i.e., pixel values) recorded by the sensors aboard satellites and (less frequently) aircraft. In most terrestrial applications of remotely-sensed data, analysts use virtually cloud-free days and/or multiple scenes to essentially avoid or ignore this issue as well as the effects of haze, cloud cover, and other atmospheric effects. However, in coastal and near-shore ocean areas, analysts must be concerned with these atmospheric effects. The following four primary methods have been developed to remove or minimize the atmospheric effects on an image (Khorram et al. in press).

Relative radiometric correction of atmospheric attenuation normalizes the intensities among different bands within a scene to remove detector-related problems, and then corrects the intensities through a comparison with a standard reference surface on the same date and same scene.

Absolute radiometric correction of atmospheric attenuation takes into account the solar zenith angle at the time of satellite overpass, the total optical thickness caused by molecular scattering, the atmospheric transmittance for a given angle of incidence, the spectral irradiance at the top of the atmosphere, and the Rayleigh and Mie scattering laws (Turner and Spencer 1972; Forster 1984; Khorram et al. 2005).

Single-image normalization uses a histogram adjustment for any shift in the histogram that may have been caused by atmospheric effects. This method is based on the fact that infrared data are largely free of atmospheric scattering effects as compared to the visible region. Thus, the histogram shifts due to haze can be used to adjust for the atmospheric effects. This method involves a subtractive bias established for each band and is very simple to use.

Multiple image normalization uses regression analysis for a number of dates. This method is used to make sure the spectral values from one date are comparable to other dates, which implicitly takes the atmospheric corrections into account. This method is primarily used for change detection purposes and is also fairly simple to use.

Radiometric Correction

Radiometric distortions vary among different sensors. Typically, solar elevation corrections and Earth-Sun distance corrections are applied to satellite data to remove the effects of the seasonal position of the Sun relative to the Earth, and to normalize for seasonal variations in the distance between the Earth and the Sun. Furthermore, noise removal algorithms can be applied to remove any malfunctions in the sensors or detectors. The causes can include signal quantization, data line drops, and recording. Several de-striping algorithms are available to remove striping and banding effects in satellite data. Line drops can be corrected by replacing the spectral values in the missing band with the average of the line(s) above and below them. Non-systematic variations in gray levels from pixel to pixel (i.e., bit errors) can be corrected by replacing these values with neighboring

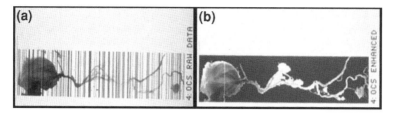

Fig. 3.3 Radiometric correction of dropped data lines in an Ocean Color Scanner image of San Pablo Bay, California, and its associated delta (Khorram 1985)

values that exceed threshold values established by the analyst, as shown in Fig. 3.3. In this example, dropped data lines that are recorded as zero values in the digital image (A) are replaced by the average of the values on the scanned lines above and below the bad data lines (B). However, when the dropped lines are more than a few pixels in width (usually 3–5 depending on the local scene dynamic), then linear averaging does not seem to be appropriate.

Band Combinations, Ratios and Indices

For simple applications such as delineating surface water boundaries, a simple process such as display of a near or middle infrared band can be sufficient. Algebraic combinations of certain bands via division, addition, subtraction, or multiplication can lead to enhanced information. Differences and ratios in various bands are primarily used for change detection and spectral enhancement studies. The most common image band ratios typically include the following: infrared band over red band for vegetation distribution; green band over red band for mapping surface water bodies and wetland delineation; red band over infrared band for mapping turbid waters; and red band over blue band or red band over green band for mineral mapping.

The most frequently used index for vegetation mapping is the Normalized Difference Vegetation Index (NDVI), which is defined as:

$$NDVI = \frac{B2 - B1}{B2 + B1}$$

where, B2 represents the brightness values (i.e., the digital numbers, or pixel values) from an infrared band of an image, and B1 represents the corresponding values in the image's red band. NDVI images can be displayed in black and white or in color, as shown in Fig. 3.4, which illustrates the global distribution of vegetation in various shades of green. NDVI images are typically used for covering large geographic areas, thus reducing the cost of data processing for certain applications such as vegetation mapping.

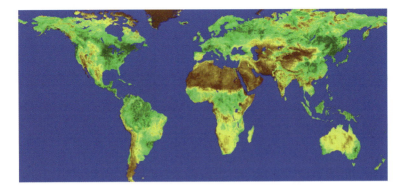

Fig. 3.4 An example NDVI image depicting global vegetation cover in various shades of green

Data Fusion

In remote sensing, **data fusion** typically refers to the process of integrating image data collected at different spatial, spectral, and/or temporal resolutions (Pohl and Van Genderen 1998; Cakir and Khorram 2008). Commonly, fusion is done at the pixel level, although it may also be performed at the feature level (i.e., recognizable objects are extracted from multiple images and then fused) or the decision level (i.e., images are processed individually and subsequently combined via decision rules) (Pohl and Van Genderen 1998). A primary objective of data fusion is to facilitate the interpretation or classification of a target geographic area in a manner, and with a degree of accuracy, that would not be possible given image data from a single source (Solberg et al. 1994; Cakir et al. 1999). This may also be seen as reducing the uncertainty associated with the initial data source.

Especially during the last two decades, fusion approaches have been advanced to combine imagery from sensors with quite disparate operational characteristics and output data formats. For instance, several studies (Simone et al. 2002; Cakir et al. 1999; Solberg et al. 1994) have outlined techniques for fusing multispectral imagery (e.g., Landsat TM, SPOT XS) and synthetic aperture radar (SAR) imagery for improved land cover classification. In addition, Lee et al. (2008) provided an example of the fusion of LiDAR and aerial imagery for automated identification and delineation of buildings. Nonetheless, one of the most common applications of data fusion in remote sensing is for **pan-sharpening** (Alparone 2008), which is the integration of a multispectral image and a corresponding, higher-resolution panchromatic image (Fig. 3.5).

A commonly used approach for data fusion is based on **principal components analysis (PCA)**. PCA is a multivariate statistical technique for reducing the dimensionality of a set of data. A comprehensive discussion of the mathematics of PCA is beyond the scope of this text, but briefly, PCA transforms an original data

Fig. 3.5 Pan-sharpening through fusion of multispectral and panchromatic Quickbird images. A true color composite (TCC) of the original multispectral image is shown at top left, while a TCC of the pan-sharpened image is shown at top right. False color composites of the original and pan-sharpened images are shown at bottom left and right, respectively (Cakir and Khorram 2008)

set (e.g., a multispectral image with four bands in the visible and near-infrared ranges) into a set of orthogonal (uncorrelated) axes or component variables. The number of these "principal" components is less than or equal to the number of variables (e.g., the number of image bands) in the original data. A critical aspect of the PCA transformation is that the first principal component explains as much of the variation in the original data as possible, while each subsequent component explains as much variation as possible while also remaining orthogonal to the previous component(s).

Although fusion is potentially quite useful, it does represent an alteration of the original data. Therefore, remote sensing analysts should carefully consider their objectives, data resolution needs, and the data and computing resources available for their projects before undertaking a fusion effort.

Image Processing

Multispectral satellite images that have been through preprocessing are then ready for processing, which essentially means they are ready for **image classification**. During classification, each pixel in an image is assigned to a particular category in a set of categories of interest, for example a set of land cover types. The classification process is generally based on the pixels' values (i.e., their spectral values) in each band of the image. A major aim of the classification process is to convert the original spectral data, which are variable and may exhibit fairly complex relationships across several image bands, into a straightforward thematic map (typically a land cover map) that is easily understood by end users. Thus, image classification is a critical step in a remote sensing project because it pulls the most essential and meaningful information out of a multidimensional data set that would otherwise be difficult to interpret. The most prominent image classification techniques utilize hard logic, meaning that image pixels are definitively assigned to a single class in the selected classification scheme. This is in contrast to fuzzy logic, which acknowledges the variability of natural phenomena and the imprecision inherent to class definitions, and so aims to characterize a pixel's probability of membership in one of possibly several classes in the scheme (Hagen 2003; Fisher and Pathirana 1990). Classification systems based on hard logic involve one of two types of classification: **supervised** or **unsupervised classification** (Hester et al. 2008). Although supervised and unsupervised classification techniques have the same end goal, they follow fundamentally different conceptual strategies. Briefly, supervised classification depends on an analyst (i.e., a human interpreter of the image) to identify **training sites** (or **signatures**) that represent each of the classes in the chosen classification scheme. Collectively, these signatures define a particular statistical profile for each class. Then analytical procedures are used to compare each pixel in the image to these statistical profiles and assign the pixel to the best matching class. Unsupervised classification involves little input from an analyst prior to the classification process. Instead, most of the work is done by an iterative computerized process that assigns pixels in an image to particular **clusters**, which represent natural groupings of pixels that are spectrally similar across the image bands. After this clustering process, the analyst then decides in which class in the selected classification scheme each cluster belongs.

Selection of a Classification Scheme

The first step in any image classification process is the selection of an appropriate **classification scheme** to denote the classes of interest to the analyst. The classes in the scheme must be defined by certain rules or criteria that distinguish them from one another and, just as importantly, must represent categories that are relevant to an end user. It is important to ensure that the selected scheme is not excessively detailed given the spatial and spectral resolution of the image being classified. The U.S. Geological Survey (USGS) classification scheme is widely used for general land cover mapping, in part because it was designed for application to remotely sensed data, and furthermore, with four levels of progressively detailed classes, offers flexibility given the input data and the end-user objectives. For classification of moderate-resolution multispectral imagery, the most frequently applied categories are from Level I and Level II (Table 3.1).

Optimum Band Selection Prior to Classification

Multispectral images have *high dimensionality* in the sense that spectral information is recorded in several bands representing different ranges of wavelengths. However, there are commonly some redundancies of the information in these bands; many recorded phenomena generate similar spectral response patterns in multiple bands. Prior to classification, especially supervised classification, it is advisable to identify a subset of bands that limits these redundancies while still retaining the most important spectral information. Feature space plots (see example in Fig. 3.6) are two-dimensional scatterplots of the pixels in an image, where each dimension represents a single image band. Each pixel is plotted as a point according to its values from the two bands. In the plot, the brightness of a given point indicates how many image pixels had the particular band value combination associated with that point. The primary aspect of interest in a feature space plot is the degree of dispersion among plot points. If the points are largely clustered along the plot diagonal, that indicates poor separability between the bands; conversely, if the points are widely dispersed in the plot space, that indicates a high degree of independence in the information recorded in the bands. Ideally, an analyst should examine feature space plots for all possible band combinations when selecting the optimal subset of bands.

Unsupervised Classification

As noted earlier, unsupervised classification methods involve limited initial input from an analyst, as a computer algorithm determines the most appropriate clusters from the image data. Nevertheless, the analyst will likely have to make certain decisions to guide the clustering algorithm, such as the number of desired output clusters, the number of iterations, and the statistical threshold values that are used to separate clusters. It is important to understand that, while most of the output clusters can be easily assigned to a particular class in the selected classification scheme, it is also common that some of the clusters will be ambiguous in that they apparently represent a mix of the classes in the scheme. The key message is that analysts should be aware that unsupervised classification, like supervised

Table 3.1 Levels I and II of the USGS Classification Scheme (modified from Anderson et al. 1976)

Level I	Level II	
1 Urban or Built-up Land	11	Residential
	12	Commercial and Services.
	13	Industrial.
	14	Transportation, Communications, and Utilities.
	15	Industrial and Commercial Complexes.
	16	Mixed Urban or Built-up Land.
	17	Other Urban or Built-up Land.
2 Agricultural Land	21	Cropland and Pasture.
	22	Orchards, Groves, Vineyards, Nurseries, and
	23	Ornamental Horticultural Areas.
	24	Confined Feeding Operations.
		Other Agricultural Land.
3 Rangeland	31	Herbaceous Rangeland.
	32	Shrub and Brush Rangeland.
	33	Mixed Rangeland.
4 Forest Land	41	Deciduous Forest Land.
	42	Evergreen Forest Land.
	43	Mixed Forest Land.
5 Water	51	Streams and Canals.
	52	Lakes.
	53	Reservoirs.
	54	Bays and Estuaries.
6 Wetland	61	Forested Wetland.
	62	Nonforested Wetland.
7 Barren Land	71	Dry Salt Flats.
	72	Beaches.
	73	Sandy Areas other than Beaches.
	74	Bare Exposed Rock.
	75	Strip Mines. Quarries, and Gravel Pits.
	76	Transitional Areas.
	77	Mixed Barren Land.
8 Tundra	81	Shrub and Brush Tundra.
	82	Herbaceous Tundra.
	83	Bare Ground Tundra.
	84	Wet Tundra.
	85	Mixed Tundra.
9 Perennial Snow or Ice	91	Perennial Snowfields.
	92	Glaciers.

classification, has limitations that impact the accuracy and interpretability of the output classified image.

Although there are many clustering methods, the two most widely used techniques for unsupervised classification of multispectral images are the Clustering

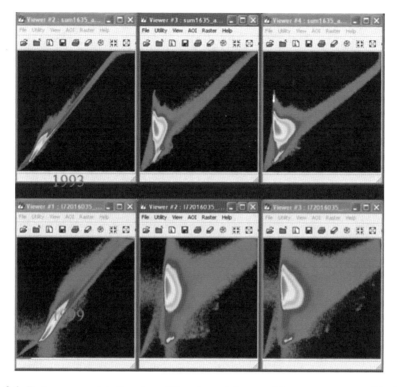

Fig. 3.6 Feature space plots from two different remotely sensed images, one captured in 1993 (*top*) and one captured in 1999 (*bottom*). A greater level of dispersion can be seen in the feature space plots for the 1999 image

Algorithms technique (also known as the chain method) and the **Iterative Self-Organizing Data Analysis Technique (ISODATA)**.

The ISODATA technique (Tou and Gonzalez 1977; Sabins 1987) is considered an iterative technique because it may make many passes through the data, rather than just two, in order to develop an adequate final set of clusters. The analyst may specify the maximum desired number of output clusters. In addition, the analyst may set the following other parameters: the maximum number of iterations, the minimum percentage of pixels assigned to each cluster, the maximum standard deviation, the minimum distance between cluster means, and most critically, the maximum percentage of pixels whose cluster labels can remain unchanged between iterations (Jensen 2005; Hester et al. 2008). This latter parameter is important because when this percentage is reached, the algorithm stops processing. Briefly, the ISODATA algorithm begins with a creation of arbitrary clusters, and in the first pass through the image, each pixel is compared to the mean vectors of these clusters and assigned to the cluster to which it is closest in spectral space. In subsequent iterations, new mean vectors are calculated for the clusters resulting from the previous iterations and the image pixels are compared to these new

Fig. 3.7 Unsupervised classification of the San Francisco Bay area, California, based on the ISODATA method

means. This process is repeated until the specified criteria are satisfied, that is until there is very little change in cluster labels between iterations or the maximum numbers of iterations has been reached (Jensen 2005). An example of an application of unsupervised classification using the ISODATA method for the San Francisco Bay area is shown in Fig. 3.7.

Supervised Classification

The training sites that an analyst selects during the initial stage of a supervised classification project serve to establish the relationships between the classes of interest and the image spectral data. To identify and delineate candidate sites, an analyst will typically have to rely on ancillary data sources, such as aerial photography, existing GIS data, or field visits. In addition, there are several rules of thumb that help ensure the training sites provide an adequate foundation for successful classification (Jensen 2005). First, the number of training sites should be at least three times the number of categories of interest. The training sites should be selected to represent the spectral distribution of each class of interest as thoroughly as possible, and should be randomly or systematically distributed throughout the image area. Finally, each training site should be as homogeneous as possible, meaning that the group of pixels in a training site should have similar spectral values. Three widely used algorithms in supervised classification are minimum-distance-to-means, parallelepiped, and **maximum likelihood**.

The **Maximum Likelihood Classification Algorithm** assumes that the training site data (i.e., the spectral values of the associated pixels) for each class are normally distributed (Blaisdell 1993). Under this normality assumption, the distributions of spectral values (from each band) for a class of interest can be validly described by a mean vector as well as a covariance matrix, which depicts the variation of values both within and between image bands. In turn, the statistical probability that any pixel in an image belongs to a particular class of interest can be calculated based on the mean vectors and covariance matrices. In short, the pixel is assigned to the class in which it has the highest probability, or maximum likelihood, of membership. Although the maximum likelihood classification algorithm is widely used, it is computationally intensive (Jensen 2005, Khorram et al. in press). Figure 3.8 illustrates an application of the maximum likelihood algorithm for image classification, as compared to a true color composite of the classified area.

A more sophisticated implementation of the maximum likelihood classifier is the Bayesian Classifier, in which weights are assigned to the probability estimates for each class based on prior knowledge of the expected likelihood of class occurrence (Hord 1982; Hester et al. 2008); for example, a historical map of the study area could provide an initial estimate of the proportion of the area represented by each class. In other words, the classes are not assumed to have equal probabilities of occurrence, as is the case with the traditional maximum likelihood classifier.

Fuzzy Logic Classification

As hard-logic classifiers, the unsupervised and supervised classification algorithms described above assign the pixels of an image to mutually exclusive categories. However, real-world conditions do not always conform to definitive boundaries between classes of interest. Because classification is done on the basis of spectral response data, and despite the best effort of an analyst to define classes uniquely, there are bound to be cases of overlap. This is especially true of low-to-moderate-spatial-resolution multispectral imagery, where image pixel values may be generated by more than one ground phenomenon, resulting in **"mixed pixels"** (Fisher and Pathirana 1990). **Fuzzy logic** classifiers circumvent the definitive boundaries of hard-logic classifiers by establishing transitional, or "fuzzy", regions between classes in a scheme, which come into play when a pixel potentially belongs to two or more classes (Jensen 2005; Zadeh 1965). For instance, for any pixel in a multispectral image, it is possible to measure the distance between the pixel's spectral values and the mean vectors (i.e., the sets of mean spectral values) for all classes in the classification scheme. These measured distances may then be straightforwardly translated to the pixel's likelihood of membership in each class.

Fig. 3.8 The results of a supervised classification using the maximum likelihood algorithm (*bottom*), as compared to a true color composite image of the targeted area

Other Classification Approaches

Many other classification approaches have been developed from research in fields such as artificial intelligence and pattern recognition. For example, **artificial neural network (ANN)** approaches have been applied for remote sensing image classification since the early 1990s (Dai and Khorram 1999; Yang and Chung 2002; Qiu and Jensen 2004). The essential processing unit of an ANN is called a neuron, which is analogous to a neuron in the human brain (Jensen et al. 2009).

Essentially, ANNs attempt to mimic the performance of the human brain for the purpose of data processing (Mohanty and Majumdar 1996). First, the ANN acquires knowledge through a learning process. Second, this knowledge is stored in the connection strengths between neurons, which are known as synaptic weights (Jensen et al. 2009; Haykin 1994). In what is known as **supervised learning**, these synaptic weights may be adapted through the application of training samples. Thus, an ANN adaptively relates inputs to outputs through its set of neurons, rather than depending on statistical procedures.

There are numerous advantages to ANNs for classification of remotely sensed images. Most prominently, they are not bound by statistical assumptions such as the assumption of normality. They are also well suited to the incorporation of ancillary data sets (e.g., GIS layers such as digital elevation models or soil maps), which is a limitation of more traditional classification algorithms such as maximum likelihood. However, the supervised learning process usually requires a great deal of computation time, and the relationships between inputs and outputs in an ANN may not be as readily discernable as with more traditional classifiers (Jensen et al. 2009).

Another useful strategy for incorporating ancillary data into image classification efforts is the implementation of an **expert system** (Stefanov et al. 2001; Goodenough et al. 1987). Essentially, an expert system is a set of decision rules that are applied to both image data and supporting geospatial data sets in order to classify each pixel in the image. The decision rules in an expert system can be applied to an unclassified image (i.e., incorporating the original spectral values in the rule set) or, instead, to an image that has already been classified using a supervised or unsupervised approach (i.e., the class assignments of pixels are used in the rule set). There are a number of statistical procedures for constructing a set of rules in order to classify multi-factor data, with decision tree methods perhaps most prominent. For examples of the use of decision tree methods in a remote sensing context, see work by Pal and Mather (2003), and Friedl and Brodley (1997).

Finally, with the advent of higher spatial resolution imagery, in the 1-meter to sub-meter range, the use of traditional algorithms have proved challenging in obtaining highly accurate classifications because the images tend to exhibit high spectral variability in the target classes (Yu et al. 2006). Newer techniques, involving **object-oriented classification** and **image segmentation algorithms** represent novel approaches to image classification (Blaschke 2010; Shackelford and Davis 2003).

The object-oriented classification procedure varies from the traditional methods in that, rather than broadly applied classifications based on groups of pixels identified as training samples, instead entire objects that make up surface features to be classified are selected (Fig. 3.9). Once these objects have been established, rules are then defined to model the object's, as well as similar objects', shape, size, distance, texture, spectral pixel values, position, orientation and relationship with other objects, as well as the distribution of the objects throughout the image, along with a number of additional user-defined parameters. Each object may represent a particular cover type. The applied object-oriented procedure then utilizes algorithms such as nearest neighbor analysis, neural networks, and decision tree analyses to complete the classification operation (Herold et al. 2003).

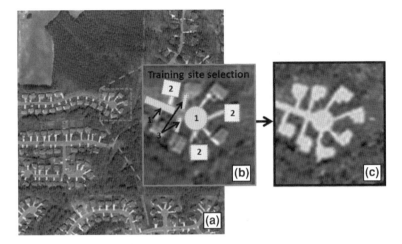

Fig. 3.9 Image (a) depicts an unclassified 2003 true color, digital orthoimage with a spatial resolution of 33 cm (USGS 2006). Inset (b) depicts three manually selected impervious surface features to be used as training samples; (1) roads, (2) houses, and (3) driveways. Inset (c) depicts the results of an object extraction classification based on the training samples of the selected impervious surface features

The object-oriented classification represents a type of image segmentation algorithm. In general, the image segmentation algorithm operates by partitioning or segmenting an image into multiple segments (or sets of pixels) based on spectral pixel values, the shape of merged pixels, and an iterative threshold limit, as well as other user-defined parameters. The segmentation process produces an image that originally represented a complex mosaic of pixels, but now represents regions of similar pixels that are easier to analyze.

Many studies have applied the object-oriented image segmentation process to studies of land use analyses, see studies by Mitri and Gitas (2004), Al Fugara et al. (2009), Miller et al. (2009).

Post-processing Image Data

A number of post-processing operations may subsequently be applied to classified images. The most typical operations involving remotely sensed data include spatial filtering, accuracy assessment, and change detection.

Spatial Filters

Spatial filters are applied to classified images in order to either highlight or downplay features in an image based on their spatial frequency. Spatial frequency is defined as the number of changes in pixel brightness value per unit distance for

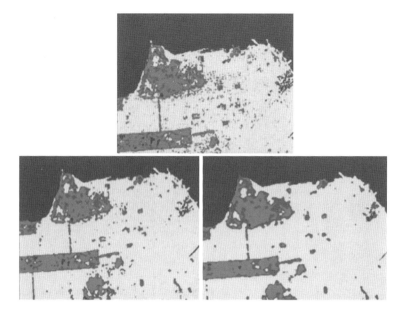

Fig. 3.10 A filtering technique has been applied to classified land use/land cover of San Francisco, CA. The original classified image (*top*) was subjected to a 3 × 3 median filter (*bottom left*) as well as a 5 × 5 median filter (*bottom right*)

any particular part of an image. High variation in brightness values translates into high spatial frequency, or what is commonly known as a salt-and-pepper effect, in classified images. Spatial filtering is the process of dividing the image into its constituent spatial frequencies, and selectively altering certain spatial frequencies to emphasize or deemphasize some image features.

Kernels, which can be thought as moving windows, are applied to classified images to reduce noise and filter unwanted information. An example of the application of a 3 × 3 and a 5 × 5 median filter to a classified map is shown in Fig. 3.10. In short, median values are calculated within a moving 3-pixel by 3-pixel or 5-pixel by 5-pixel window and assigned to the central pixel. Mean and median filters usually have a smoothing effect. Other filters that have been frequently used in image processing employ functions that highlight differences between pixel values. They are usually used for sharpening the edges of objects in an image. These edge detection filters are also known as directional filters.

Accuracy Assessment

Accuracy assessment has been a key component and the focus of a significant number of remote sensing studies (Van Genderen and Lock 1978; Congalton 1991; Goodchild et al. 1992; Khorram et al, 1992, 1999; Congalton and Green 1999).

Table 3.2 Change matrix of area (expressed in percentage values) of land converted from one land use/land cover type to the next between 1980 (Time Step 1) and 2010 (Time Step 2) as seen in Fig. 3.12[*]

		2010					
		Forest	Agriculture	Barren land	Urban land	Water	**1980 Total**
1980	Forest	(37.5)	0	0	6.25	0	**43.75**
	Agriculture	0	(12.5)	0	25	0	**37.5**
	Barren Land	0	0	(0)	6.25	0	**6.25**
	Urban Land	0	0	0	(6.25)	0	**6.25**
	Water	0	0	0	0	(6.25)	**6.25**
	2010 Total	**37.5**	**12.5**	**0**	**43.75**	**6.25**	**100**

[*] Changes here show the total percentages of each land use/land cover category that changed into another category. For example, 18.75% total area was converted from agriculture to urban land from 1980 to 2010. This change represents a "*From-To*" land conversion. The total column represents the land use/land cover areas in the earlier time step (1980). The areas of no change are in parentheses

Without assessing the accuracy of a classified data, the reliability and repeatability of the output products are in question.

The most commonly used procedure for accuracy assessment is error matrix analysis. As shown in Table 3.2, an error matrix can be constructed from the results based on reference data (e.g., data collected on the ground or from a substantially higher spatial resolution image) on one side of a table, and the results based on the classified image on the other side of the table. An adequate number of samples are identified on the classified image and corresponding reference data are collected using various sampling strategies. The accuracy is determined in terms of percent correctly classified sample sites, as compared to their corresponding reference data, for each category of interest as well as the overall classification accuracy involving all categories. Traditionally, the total number of correct samples in a given category is divided by the total number of samples in that category based on reference data. This accuracy measure indicates omission errors and is often referred to as "producer's accuracy" because the producer of the image classification is interested in how well he has classified a certain category. If the total number of correctly-identified samples in a given category is divided by the total number of samples based on classified data, then this indicates the commission error. This measure is called the "user's accuracy" or reliability because the user is interested in the probability that a classified sample representing the actual category on the ground. Multivariate statistical procedures have also been used for accuracy assessment. The most commonly used is a discrete multivariate technique, called KAPPA, which is a measure of agreement or accuracy by KHAT statistics (Cohen 1960).

Change Detection

Change detection using remote sensing data is an attempt to record natural and anthropogenic transitions of land use/land cover (LU/LC) on Earth's surface that

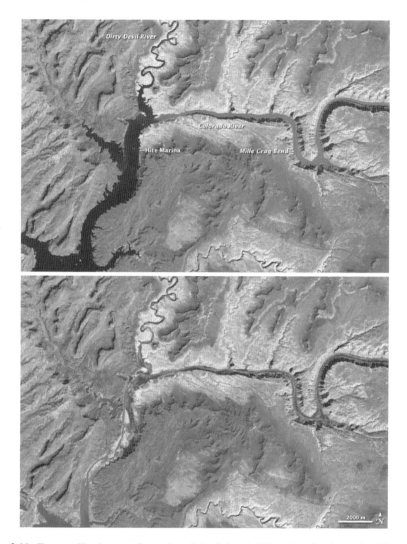

Fig. 3.11 Two satellite images of a section of the Colorado River, showing dramatic differences in water levels when the images were collected. Delineation and quantification of such differences are common objectives of change detection

have occurred over time (e.g., see Fig. 3.11). This type of analysis is usually completed by using either manual or automated methods, or a combination of both (Nelson et al. 2002; Cakir et al. 2006; Hester et al. 2010). Automated methods can be categorized as pre-classification, post-classification, or even constitute more advanced procedures.

Pre-classification methods develop change maps from multi-temporal data (i.e. data captured over the same area on different dates) without first generating classified LU/LC maps from that data. The algorithms used in pre-classification

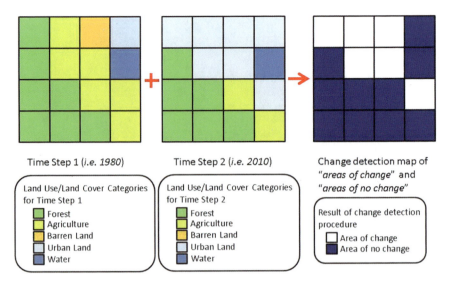

Fig. 3.12 Depicts the change occurring between Time Step 1 and Time Step 2 (i.e. 30 years change from 1980–2010)

procedures may transform or simplify the original data before creating a change map, but they do not rely first on the generation of meaningful of LU/LC classifications of the individual image dates. One of the most important aspects of any pre-classification change detection algorithm is the specification of a change threshold. This parameter represents the interpretative mechanism by which the algorithm judges whether a change has occurred.

Some pre-classification methods that rely on user-defined change thresholds include image differencing, image regression, imaging ratioing, vegetation index differencing, principal component analysis (PCA), and change vector analysis (Lu et al. 2004). If input imagery is suitably matched in terms of geometric, radiometric, and atmospheric quality, then the major advantage of pre-classification change detection is that these approaches do not introduce a great deal of analyst subjectivity as they evaluate the quantitative differences in image dates, despite being computationally intensive. One limitation to pre-classification change analysis is that geometric and radiometric alignment between image dates may be difficult or even impossible to achieve. This limits the ability to develop a complete "*from-to*" representation of change (Jensen 2005). Pre-classification methods are primarily equipped to identify either binary or scaled indicators of "*change*" or "*no change*" (Fig. 3.12).

Post-classification change detection can generally provide a detailed matrix of "*from-to*" change and requires a great deal of preparatory analyst input (Table 3.2). In this approach, the original study imagery (image dates 1 and 2, or more) is thematically classified. The change analysis occurs when the classified maps are compared by one or more of the diverse post-classification methods.

These include the expectation-maximization algorithm, artificial neural networks, unsupervised change detection, and hybrid change detection (Lu et al. 2004). In addition to the advantage of the *"from-to"* output generated by these approaches, post-classification change detection can minimize the effect of atmospheric and environmental disparity between input images. However, it is important for the analyst to be aware that each classification input introduces error into the post-classification change analysis, and this error is magnified in the final change map by the aggregate classification error from each input image.

More advanced change analysis methods integrate ancillary GIS data, statistical models, spatial analysis, textural metrics, or other alternatives to pre- and post-classification approaches (Lu et al. 2004). These methods are too diverse to befit many general characterizations, but they are usually developed in response to specific research area issues or LU/LC categories of particular interest. Examples of advanced methods include those using measures of spatial dependence (Henebry 1993), generalized linear models (Morisette et al. 1999), spatial-spectral structure (Zhang et al. 2002), and spatial statistics (Read and Lam 2002).

References

A.M. Al Fugara, B. Pradhan, T.A. Mohamed, Improvement of land-use classification using object-oriented and fuzzy logic approach. Appl. Geomat. **1**, 111–120 (2009)

L. Alparone, B. Alazzi, S. Baronti, A. Garzelli, F. Nencini, M. Selva, Multispectral and panchromatic data fusion assessment without reference. Photogramm. Eng. Remote Sens. **74**(2), 193–200 (2008)

J.R. Anderson, E. Hardy, J. Roach, R. Witmer, A land use and land cover classification system for use with remote sensing data (US Geological Survey Professional Paper 964, Washington, 1976), p. 28

E.A. Blaisdell, *Statistics in Practice* (Harcourt Brace Jovanovich, New York, 1993), p. 653

T. Blaschke, Object based image analysis for remote sensing. ISPRS J. Photogramm. Remote Sens. **65**, 2–16 (2010)

H.I. Cakir, S. Khorram, Pixel level fusion of panchromatic and multispectral images based on correspondence analysis. Photogramm. Eng. Remote Sens. **74**(2), 183–192 (2008)

H.I. Cakir, S. Khorram, X.L. Dai, P. de Fraipont, Merging SPOT XS and SAR imagery using the wavelet transform method to improve classification accuracy, in *Proceedings of the IEEE 1999 International Geoscience and Remote Sensing Symposium* (IGARSS28 June–2 July, Hamburg, 1999), pp. 71–73

H.I. Cakir, S. Khorram, S.A.C. Nelson, Correspondence analysis approach for detecting land use land cover changes. Remote Sens. Environ. **102**, 306–317 (2006)

J.A. Cohen, A coefficient of agreement for nominal scales. Educational. Psychol. Meas. **20**, 37–46 (1960)

R.G. Congalton, A review of assessing the accuracy of classifications of remotely sensed data. Remote Sens. Environ. **37**, 35–46 (1991)

R.G. Congalton, K. Green, *Assessing the Accuracy of Remotely Sensed Data: Principles and Practices* (Lewis Publishers, Boca Raton, 1999), p. 137

X. Dai, S. Khorram, Data fusion using artificial neural networks: a case study on multitemporal change analysis. Comput. Environ. Urban. Syst. **23**, 19–31 (1999)

P.F. Fisher, S. Pathirana, The evaluation of fuzzy membership of land cover classes in the suburban zone. Remote Sens. Environ. **34**, 121–132 (1990)

B.C. Forster, Derivation of atmospheric correction procedures for Landsat MSS with particular reference to urban data. Int. J. Remote Sens. **5**, 799–817 (1984)

M.A. Friedl, C.E. Brodley, Decision tree classification of land cover from remotely sensed data. Remote Sens. Environ. **61**, 399–409 (1997)

M.F. Goodchild, G.Q. Sun, S. Yang, Development and test of an error model for categorical data. Int. J. Geogr. Inf. Syst. **6**(2), 87–104 (1992)

D.G. Goodenough, M. Goldberg, G. Plunkett, J. Zelek, An expert system for remote sensing. IEEE. Trans. Geosci. Remote Sens. **GE-25**, 349–359 (1987)

A. Hagen, Fuzzy set approach to assessing similarity of categorical maps. Int. J. Geogr. Inf. Sci. **17**, 235–249 (2003)

S. Haykin, *Neural Networks: A Comprehensive Foundation* (Prentice Hall, Englewood Cliffs, 1994)

G.M. Henebry, Detecting change in grasslands using measures of spatial dependence with Landsat TM data. Remote Sens. Environ. **46**, 223–234 (1993)

M. Herold, X.H. Liu, K.C. Clarke, Spatial metrics and image texture for mapping urban land use. Photogramm. Eng. Remote Sens. **69**, 991–1001 (2003)

D.B. Hester, Land cover mapping and change detection in urban watersheds using quickbird high spatial resolution satellite imagery, Ph.D. Dissertation, North Carolina State University, Raleigh (2008), p. 148

D.B. Hester, H.I. Cakir, S.A.C. Nelson, S. Khorram, Per-pixel classification of high spatial resolution satellite imagery for urban land cover mapping. Photogramm. Eng. Remote Sens. **74**, 463–471 (2008)

D.B. Hester, S.A.C. Nelson, H.I. Cakir, S. Khorram, H. Cheshire, High-resolution land cover change detection based on fuzzy uncertainty analysis and change reasoning. Int. J. Remote Sens. **31**, 455–475 (2010)

R.M. Hord, *Digital Image Processing of Remotely-Sensed Data* (Academic Press, New York, 1982), p. 256

J.R. Jensen, *Introductory Digital Image Processing*, 3rd edn. (Pearson Prentice Hall, Upper Saddle River, 2005), p. 316

R.R. Jensen, P.J. Hardin, G. Yu, Artificial neural networks and remote sensing. Geogr. Compass **3**(2), 630–646 (2009)

S. Khorram, S.A.C. Nelson, H.I. Cakir, C. Van Der Wiele, Digital image processing, post-processing, and data integration, *Handbook of Satellite Applications*, 1st ed., ed. by J.N. Pelton, S. Madry, S. Camacho-Lara (Springer-Verlag, New York, in press)

S. Khorram, Development of water quality models applicable throughout the entire San Francisco Bay and delta. Photogramm. Eng. Remote Sens. **51**(1), 53–62 (1985)

S. Khorram, S.N. Nelson, H.I. Cakir, D.B. Hester, H.M. Cheshire, Cost effective assessment of land use practices influencing erosion and sediment yield, Technical report, Center for Earth Observation, North Carolina State University, submitted to North Carolina Water Resources research Institute (2005), p. 46

S. Khorram, G.S. Biging, N.R. Chrisman, D.R. Colby, R.G. Congalton, J.E. Dobson, R.L. Ferguson, M.F. Goodchild, J.R. Jensen, T.H. Mace, *Accuracy Assessment of Remote Sensing-Derived Change Detection* (American Society of Photogrammetry and Remote Sensing Monograph, Bethesda, 1999), p. 64

S. Khorram, H.M. Cheshire, K. Sidrellis, Z. Nagy, *Mapping and GIS Development of Land Use/Land Cover Categories for the Albemarle-Pamlico Drainage Basin*. North Carolina Department of Environmental, Health, and Natural Resources, Dept. No. 91-08 (1992), p. 55

D.H. Lee, K.M. Lee, S.U. Lee, Fusion of LiDAR and imagery for reliable building extraction. Photogramm. Eng. Remote Sens. **74**, 215–225 (2008)

D. Lu, P. Mausel, E. Brondizio, E. Moran, Change detection techniques. Int. J. Remote Sens. **25**, 2365–2407 (2004)

J.E. Miller, S.A.C. Nelson, G.R. Hess, An object extraction approach for impervious surface classification with very-high-resolution imagery. Prof. Geogr. **61**, 250–264 (2009)

G.H. Mitri, I.Z. Gitas, A performance evaluation of a burned area object-based classification model when applied to topographically and nontopographically connected TM imagery. Int. J. Remote Sens. **25**, 2863–2870 (2004)

K.K. Mohanty, T.J. Majumdar, An artificial neural network (ANN) based software package for classification of remotely sensed data. Comput. Geosci. **22**, 81–87 (1996)

J.T. Morisette, S. Khorram, T. Mace, Land-cover change detection enhanced with generalized linear models. Int. J. Remote Sens. **20**, 2703–2721 (1999)

S.A.C. Nelson, P.A. Soranno, J. Qi, Land cover change in the Upper Barataria Basin Estuary, Louisiana, from 1972–1992: increases in wetland area. Environ. Manag. **29**, 716–727 (2002)

M. Pal, P.M. Mather, An assessment of the effectiveness of decision tree methods for land cover classification. Remote Sens. Environ. **86**, 554–565 (2003)

C. Pohl, J.L. Van Genderen, Multisensor image fusion in remote sensing: concepts, methods and applications. Int. J. Remote Sens. **19**(5), 823–854 (1998)

F. Qiu, J.R. Jensen, Opening the black box of neural networks for remote sensing image classification. Int. J. Remote Sens. **9**, 1749–1768 (2004)

J.M. Read, N.S.N. Lam, Spatial methods for characterising land cover and detecting land-cover changes for the tropics. Int. J. Remote Sens. **23**, 2457–2474 (2002)

M.J. Sabins, Convergence and consistency of fuzzy c-means/ISODATA algorithms. IEEE Trans. Pattern. Anal. Mach. Intell. **9**, 661–668 (1987)

A.K. Shackelford, C.H. Davis, A combined fuzzy pixel-based and object-based approach for classification of high-resolution multispectral data over urban area. IEEE. Trans. Geosci. Remote Sens. **41**, 2354–2363 (2003)

G. Simone, A. Farina, F.C. Morabito, S.B. Serpico, L. Bruzzone, Image fusion techniques for remote sensing applications. Inf. Fusion **3**, 3–15 (2002)

A.H.S. Solberg, T. Taxt, A.K. Jain, Multisource classification of remotely sensed data: fusion of Landsat TM and SAR images. IEEE Trans. Geosci. Remote Sens. **32**(4), 768–778 (1994)

W.L. Stefanov, M.S. Ramsey, P.R. Christensen, Monitoring urban land cover change: an expert system approach to land cover classification of semiarid to arid urban centers. Remote Sens. Environ. **77**, 173–185 (2001)

J.T. Tou, R.C. Gonzalez, *Pattern Recognition Principles* (Addison-Wesley, Reading, 1977), p. 377

R.E. Turner, M.M. Spencer, Atmospheric model for correction of spacecraft data, in *Proceedings of the Eighth Annual Symposium on Remote Sensing of Environment*, ERIM, Ann Arbor (1972), pp. 895–934

U.S. Geological Survey (USGS) EO-1 Home—earth observing mission 1 (2006), http://eo1.gsfc.nasa.gov/. Accessed 27 May 2011

J.L. Van Genderen, B.F. Lock, P.A. Vass, Remote sensing: statistical testing of map accuracy. Remote Sens. Environ. **7**, 3–14 (1978)

C. Yang, P. Chung, Knowledge-based automatic change detection positioning system for complex heterogeneous environments. J. Intell. Robotic Syst. **33**, 85–98 (2002)

Q. Yu, P. Gong, N. Clinton, G. Biging, M. Kelly, D. Schirokauer, Object-based detailed vegetation classification with airborne high spatial resolution remote sensing imagery. Photogramm. Eng. Remote Sens. **72**, 799–811 (2006)

L.A. Zadeh, Fuzzy sets. Inf. Control **8**(3), 338–353 (1965)

Q. Zhang, J. Wang, X. Peng, P. Gong, P. Shi, Urban built-up land change detection with road density and spectral information from multi-temporal Landsat TM data. Int. J. Remote Sens. **23**, 3057–3078 (2002)

Suggested Reading

X. Dai, S. Khorram, A new automated land cover change detection system for remotely-sensed imagery based on artifical neural networks, in *Proceedings of IEEE/IGARSS'97 International Geoscience and Remote Sensing Symposium*, Singapore (1997)

A.K. Jain, *Fundamentals of Digital Image Processing* (Prentice Hall, Englewood Cliffs, 1989), pp. 418–421

T. Lillesand, R. Kiefer, J. Chipman, *Remote Sensing and Image Interpretation*, 6th edn. (Wiley, New York, 2008), p. 763

R.L. Lunetta, J.G. Lyons (eds.), Geospatial Data Accuracy Assessment, Report No. EPA/600/R-03/064 (US Environmental Protection Agency, Las Vegas, 2003), p. 335

Photogrammetric Engineering and Remote Sensing, Special Issue on Remote Sensing Data Fusion, Vol. **74**(2), (Feb 2008)

Chapter 4
Using Remote Sensing
for Terrestrial Applications

Every day, literally millions of individual images and observations are collected, allowing the ability to examine, monitor, and model ecosystem health, assess atmospheric composition, detect seismic activity, identify surface vegetation, perform agricultural monitoring, reveal polar ice fluctuations, expose humanitarian violations, and obtain an enormous variety of information about the earth's surface and subsurface.

The evolution of satellite remote sensing and geospatial technology over the past several decades has dramatically augmented our understanding of environmental phenomena, allowing us to develop innovative solutions to ever more complex social, environmental, and economic challenges. The decreasing cost of computing technology and of satellite imagery; the robust development of proprietary and open source software; the rapid growth of available geospatial data (e.g., online purchasing); the increasing image resolutions (e.g., under one meter); and the increasing number of satellite sensors imaging the earth have all contributed to transformational changes in every field of study and professional practice. This chapter provides a number of case study examples of the diverse and exciting ways that remote sensing has been applied to difficult local, regional, and global challenges and issues around the world.

Terrestrial applications of remote sensing involve mapping, change detection, monitoring, modeling, and other observations on the land use due to human activities and land cover based on what naturally exist on the earth's surface at many scales. Following are examples of the most common applications, but does not represent an exhaustive list.

Land Use and Land Cover

Classifying remotely sensed images into land use and land cover (LU/LC) categorizes natural and human made features, allowing users to examine and document landscapes for a variety of studies and applications. Resource managers and researchers use land use/land cover maps to study changes in plant composition and human settlement patterns to answer questions such as, "how is vegetation changing and why"? Several

S. Khorram et al., *Remote Sensing*, SpringerBriefs in Space Development,
DOI: 10.1007/978-1-4614-3103-9_4, © Siamak Khorram 2012

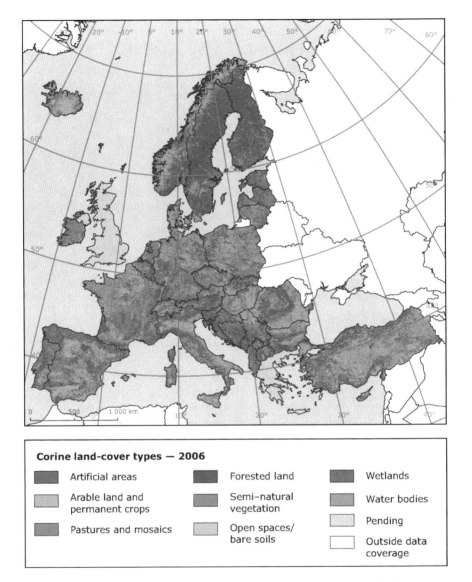

Fig. 4.1 The European Environmental Agency's CORINE 44-category land use land cover map of European countries. Image courtesy of EEA

large-scale LU/LC studies include the European Environmental Agency's classification of 25 EC Member States and other European Countries (Fig. 4.1).

LU/LC maps are commonly used in urban and regional planning, environmental studies for new infrastructure projects (e.g., highways, pipelines, etc.). Land use and land cover applications are one of the most widespread uses of remotely-sensed images.

Water Resources

There are a wide variety of ways in which remotely-sensed images are used for water resources investigations. Surface water supply reservoir mapping and monitoring, water quality mapping and modeling, runoff forecast modeling, drainage network mapping, hydrologic research, watershed characterization, and surface and ground water pollution all rely on satellite images.

Changes in land use and land cover may have severe impacts on aquatic ecosystems. Activities such as agriculture, urbanization, and erosion are primary sources of sediment and nutrient input into water bodies through surface or groundwater (Feierabend and Zelazny 1987; Dahl 1990). A few examples of the consequences of increased sediment and nutrient loading on waterbodies include the loss of littoral or coastal zone habitat, eutrophication, depletion of oxygen, and changes in food webs dynamics (Petersen et al. 1991; Carpenter et al. 1998; Wear et al. 1998).

Nonpoint source pollution (NPS) is the term used to describe loadings to water bodies from diffuse sources within the landscape. These sources make up nutrient and chemical pollutants from developed areas, construction sites, paved surfaces, agricultural and pastures lands, and atmospheric deposition. However, despite the variety of pollutants classified as NPS pollution, sediments are the largest amounts of material transported to surface waters (Loehr 1974, Tim and Jolly 1994). Approximately 60% of all sediment and sediment-associated contaminants delivered to the United States' rivers, lakes, and coastal regions are from agricultural soil loss (Madden et al. 1987; Petersen et al. 1991).

Remote sensing provides a unique technological advantage in monitoring aquatic systems due to the sensor's ability to capture data over large areas, and assorted spatial and temporal resolutions. In particular, remote sensing has been identified as an effective means of detecting and monitoring land use and land cover change in the terrestrial environment (Green et al. 1994). However, satellite remote sensing of aquatic systems has been less studied as a result of the difficulties inherent in interpreting reflectance values of water (Verbyla 1995). Clear water provides little spectral reflectance because longer wavelengths are absorbed and the reflected shorter wavelengths, which sensors rely on for surface feature detection, are subject to atmospheric scattering. Due to the absorption of longer wavelengths, most deep and clear water bodies appear dark on unclassified satellite imagery (Fig. 4.2). Although sensors such as the Landsat Multispectral Scanner (MSS) and Thematic Mapper (TM) were primarily designed for land studies, and may not be completely applicable to aquatic studies (Nelson et al. 2006). Recent improvements in sensor technologies now provide better spatial and spectral resolutions than previously available, which may improve their usefulness to aquatic studies (Kloiber et al. 2000).

In a study by Nelson et al. (2003), Landsat-7 data was used to investigate the relationship between spectral values extracted from Landsat Enhanced Thematic Mapper (ETM+) data and water clarity using Secchi disk transparency (SDT) measurements (Fig. 4.3).

Fig. 4.2 Due to the
absorption of longer
wavelengths, most deep and
clear water bodies appear
dark on unclassified satellite
imagery

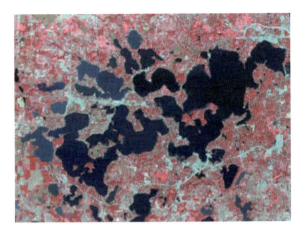

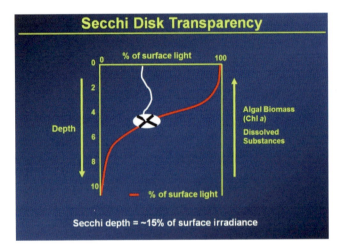

Fig. 4.3 Secchi disk transparency has been used as a simple method of determining water clarity. The Secchi depth is inversely proportional to absorbance of light by water and dissolved or particulate substances, such as algal biomass

Models that incorporate this complete range of SDT values may not produce statistical relationship results as high as models that are developed for groups of lakes that don't incorporate the full regional range of SDT values (Lathrop and Lillesand 1986, Dekker and Peters 1993, Kloiber et al. 2000). In addition, Landsat may be better suited for predicting shallow SDT values (eutrophic-higher algal biomass) than clearer water containing deeper SDT values (Fig. 4.4). These types of models, based on remotely sensed data, have the capacity to provide managers with a method to inventory shallow SDT lakes across the entire state where in situ data is difficult to obtain. In particular, regulatory and management agencies would be able to monitor private lakes without spending resources on these lakes and

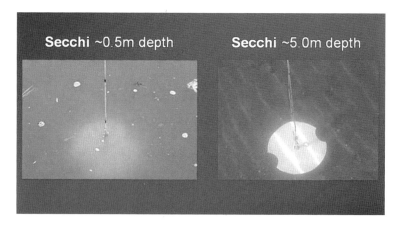

Fig. 4.4 The physical appearance of a water body with a shallow Secchi Disk Transparency (SDT) value vs deep SDT value can be very different. The image on the *left* represents a shallow SDT measurement where the water may be discolored dues to high algal biomass content, sediment, or etc. The image on the *right* represents a deep SDT measurement where the water appears clearer. Thus, the measurement of SDT values can provide important information as an indication of water clarity

without requiring the permission of the property owner. Additionally, models developed from remotely sensed data may be useful in the historical analysis of watershed changes that may have caused lake eutrophication to occur over time. This application is a very attractive use of remote sensing since historical imagery may be obtained for aquatic systems that have become eutrophic over time.

Satellite and aircraft data has also been successfully applied for modeling water quality parameters such as turbidity, chlorophyll concentrations and suspended solids (Khorram 1985; Khorram and Cheshire 1985; Catts et al. 1985). Figure 4.5 is an example of chlorophyll-*a* concentrations over the San Francisco Bay, California, derived from Landsat TM data. A typical process involves collecting surface water quality samples simultaneously with satellite or aircraft data acquisition. A regression model is then developed between a portion of the water quality sample data and the digital numbers in each of the satellite or aircraft data corresponding to the location of each sample site. These models are then verified using the other portion of the water sample data not utilized for the development of the models (Khorram 1985).

Despite current limitations, remote sensing does possess the potential to provide a valuable tool for monitoring freshwater systems. Several studies have successfully used remote sensing to measure characteristics of aquatic systems such as chlorophyll and Secchi disk transparency, suspended sediments and dissolved organic matter (Lathrop and Lillesand 1986; Jensen et al. 1993; Narumalani et al. 1997; Lillesand et al. 1983; Khorram and Cheshire 1985; Dekker and Peters 1993; Kloiber et al. 2000; Nelson et al. 2003). Other studies have used remote sensing to map aquatic plant abundance in large areas of homogenous cover or for species detection in smaller areas they have used high resolution hand or aircraft sensors (Ackleson and Klemas 1987; Armstrong 1993, Penuelas et al. 1993; Lehmann and Lachavanne 1997; Nelson

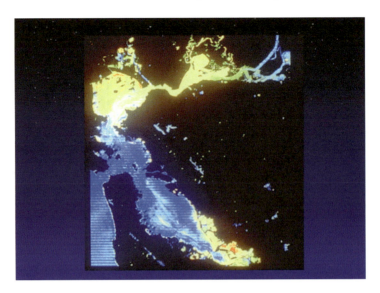

Fig. 4.5 Chlorophyll-*a* concentrations in San Francisco Bay. The values increase as the colors change from *blue* to *green* to *yellow* to *orange* to *red*

et al. 2006). While remote sensing has been used in large-scale land change studies for many years, few studies have used currently available remotely sensed data to develop standardized methods of measuring water quality parameters across a wide range of aquatic environments, over large regions, or covering multiple images.

Accessing water quality parameters on a regional scale provides a challenge in that regional remote sensing requires that multiple remotely sensed images be combined from multiple dates or regions of varying atmospheric influences (i.e. haze, cloud cover, etc.). Multi-temporal image-to-image differences are a result of atmospheric absorption and light scatter, which can be highly variable from one period of time to the next (Moore 1980). This difference can have a pronounced effect on the interpretation of imagery for trend analyses when using multiple images that are collected on different days, months, or years. Song et al. (2001) found that atmosphere corrections are necessary when using two or more satellite images collected on different dates, such as imagery used in land use/cover change studies. For example, for aquatic studies, Brivio et al. (2001) found that without the appropriate corrections for atmospheric scattering and transmittance effects, successful remote sensing of water quality parameters could not be accurately achieved from four Landsat TM images. Brivio et al. (2001) were able to develop two image-based rectification models which were used to correct regional atmospheric effects in multi-seasoned Landsat images of Lake Lseo and Lake Garda, Italy. However, accurate and simple procedures to correct for these effects are still an area of future research (Rahman 2001; Song et al. 2001).

These atmospheric limitations may be reduced with many recent technological advances. A few of the newest sensors include IKONOS, the EOS (Earth Observing System) Terra sensors, and the Landsat ETM+. These sensors will ultimately employ more advanced data calibrations than possible with some of the older sensors. However, the advantage of the older Landsat platforms is that they provide consistent historical data not yet available from new sensors and the temporal and spatial resolutions provided by Landsat MSS and TM are still more effective in monitoring landscape changes.

The availability of newer data will, however, allow for the integration of cross-platform data in developing band and image sharpening algorithms which will serve to fill in the gaps present in the older system data acquisitions (clouds, haze, etc.) and enhance the ability to develop multi-image calibrations for cross referencing. Furthermore, three of the main Terra sensors (MODIS—Moderate Resolution Imaging Spectrometer, ASTER—Advanced Spaceborne Thermal Emission and Reflectance Radiometer, and MISR—Multi-angle Imaging Spectroradiometer) have been identified as being particularly useful in the remote sensing of large area land cover and land use changes and vegetation dynamics studies (NASA 1998; Jensen 2000). However, aquatic remote sensors have yet to see the production of sensors specifically designed to monitor aquatic environments.

Currently, Landsat provides one of our richest data sources for the remote sensing of historical changes in large areas of wetlands, water quality and aquatic plants. This is evident by the high revisit cycle, spatial resolutions, and 30 years of operation provided by these sensors.

Forest Resources

Applications of remotely-sensed data to forest resources and environmental management include: forest canopy extent and structure mapping, forest health mapping and monitoring, timber inventory, timber stand conditions assessment, harvesting and procurement priority implementation, reforestation, and of course, deforestation.

Deforestation, the clear-cutting of forests for subsequent conversion to other land uses, is common throughout much of the world, especially in tropical regions (Asner et al. 2006; Curran et al. 2004; Curran and Trigg 2006; Hansen et al. 2008). During the past decade, the five countries with the highest rates of forest loss (Brazil, Papua New Guinea, Gabon, Indonesia, and Peru) lost a combined 3.6 million ha of primary forest each year (Koh et al. 2011). Large-scale deforestation has major implications for ecosystem function and biodiversity, as well as climate change, since it affects regional carbon storage and sequestration (Hansen et al. 2010; Kuemmerle et al. 2009; Laporte et al. 2007; Linkie et al. 2004). In addition, selective logging is widespread in rapidly deforesting areas such as the Amazon, further exacerbating the ecological impacts (Laporte et al. 2007, Asner et al. 2006).

Deforestation is typically driven by economic pressure (i.e., a perceived economic incentive for forest conversion rather than conservation) and is usually facilitated by limited governmental oversight in the affected regions (Koh et al. 2011; Kuemmerle et al. 2009; Laporte et al. 2007; Curran et al. 2004). Because of the lack of regulatory enforcement, illegal logging is often rampant, encroaching upon (and reaching into) supposedly protected forest areas (Curran et al. 2004; Linkie et al. 2004).

Unfortunately, governments in those world regions with the highest deforestation rates typically lack resources for monitoring or predicting patterns of forest conversion (Linkie et al. 2004). However, a number of recent studies have shown that remote sensing can be a cost-effective approach for monitoring deforestation (due to both legal and illegal logging) and for identifying management priority areas. The typical remote-sensing-based strategy is to analyze a time series of Landsat images (or similar moderate-resolution imagery) captured over a period of many years (Curran and Trigg 2006). Time series analyses of moderate-resolution satellite images have been used to study deforestation in many regions of the world: Southeast Asia (Koh et al. 2011; Curran et al. 2004; Linkie et al. 2004), the Amazon (Asner et al. 2006), Central Africa (Laporte et al. 2007), and Eastern Europe (Kuemmerle et al. 2009). Below, we highlight a few such applications.

Instead of focusing on actual changes in forest cover, Laporte et al. (2007) mapped the expansion of logging roads over a 27-year period (1976–2003) in the humid forests of Central Africa. Despite the persistence of cloud cover in this region (an estimated 1.6% of which was completely obscured by cloud cover), the researchers typically had multiple images available to delineate an area; in total, they used more than 300 Landsat images to map their 4 million km^2 study area. The images were contrast stretched to optimize road detection, and then the roads were manually digitized. Basically, the researchers found that this manual approach outperformed semi-automated approaches, particularly when delineating older roads or working from older imagery. The researchers also used IKONOS imagery of a 25 km^2 area in the northern Republic of Congo to examine the finer-scale impacts of logging activities, finding that canopy gaps in logged areas were five to six times larger than gaps in adjacent unlogged areas.

Laporte et al. (2007) mapped nearly 52,000 km of road in the study region (Fig. 4.6). Overall, logging roads accounted for 38% of the length of all roads. The most rapidly changing region was the northern Republic of Congo, where the rate of road construction increased from 156 km per year for 1976–1990 to 660 km per year after 2000. The researchers estimated that approximately 5% of the total forested area was disturbed. In summary, this effort yielded the first comprehensive data set on the status of logging in Central Africa, and so will serve as a key resource for future planning and management.

Deforestation in the Amazon region of South America is a well-known problem, but actually less than 1% of the region was deforested prior to 1975 (Guild et al. 2004). Deforestation rates increased exponentially in subsequent decades because of economic incentives for conversion to agriculture (e.g., favorable credit policies for cattle ranchers) (Guild et al. 2004, Moran 1993). The state of Rondônia in western Brazil has become one of the most rapidly deforested parts of the Amazon (Pedlowski et al. 2005; Guild et al. 2004). Many migrants have been brought to the state through government colonization projects (Pedlowski et al. 2005), placing a huge burden on the state's remaining forests. While only 4,200 km^2 of forest had been cleared in Rondônia in 1978, by 2003 the amount of cleared forest had increased to nearly 68,000 km^2 (Pedlowski et al. 2005).

Figure 4.7 shows 2000 and 2010 MODIS true color composite images from the remote northwestern corner of Rondônia. The time series illustrates the typical

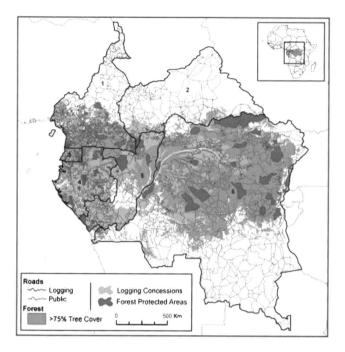

Fig. 4.6 Logging concessions and road distribution in Central Africa: Cameroon *1* Central African Republic *2* Equatorial Guinea *3* Gabon *4* Republic of Congo *5* Democratic Republic of Congo *6*. Image credit: Woods Hole Research Center (www.whrc.org)

pattern of initial deforestation along (both legal and illegal) roads, which permits small farmers to colonize previously inaccessible areas (NASA 2011). Over time, this leads to the establishment of large settlements and the clearing of most forest parcels.

Like the state of Rondônia, the southeastern Asian nation of Papua New Guinea has similarly seen a large population increase in recent decades, combined with an increased demand for timber and other resources (Shearman et al. 2009). A team of researchers from the University of Papua New Guinea and Australian National University analyzed deforestation and forest degradation in the nation—home to the world's third largest rain forest (BBC 2008)—over a 30-year period. They used Landsat, Spot 4, and Spot 5 imagery to create a map of land cover as of 2002, which they then compared to a 1972 land cover map created by the Australian Army (Shearman et al. 2009). The researchers found that 15 percent of Papua New Guinea's forest had been cleared over the 30-year period, while an additional nine percent was severely degraded by logging. They also found that the rate of forest conversion had generally risen since 1990; by 2001, the estimated deforestation rate was just over 3,600 km^2 per year (Shearman et al. 2009; BBC 2008). The current rate of conversion could result in the loss of more than half the nation's forest cover by 2021 (BBC 2008).

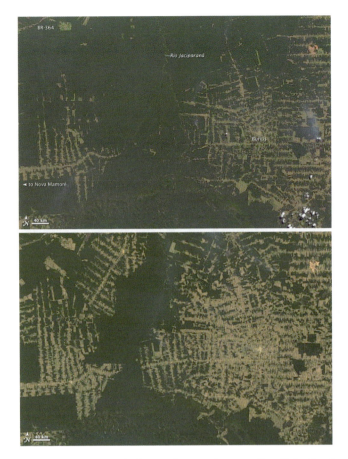

Fig. 4.7 MODIS true color composite images from northwestern Rondônia, Brazil, captured in 2000 (*top*) and 2010 (*bottom*). Images courtesy of NASA

Figure 4.8 shows two satellite images from Papua New Guinea's Gulf Province, the first captured in 1988 and the second in 2002. Because Papua New Guinea is a mountainous island country, logging has mostly been limited to accessible areas (i.e., valleys and coastal lowlands). However, Shearman et al. (2009) estimated that 13% of upper montane forests were also lost between 1972 and 2002.

Agricultural Applications

Satellite remote sensing has been increasingly applied to '**precision agriculture**' around the world. Forecasting and improving production levels results in: (1) price stability, (2) optimizes the use of storage, transport, and processing facilities, and

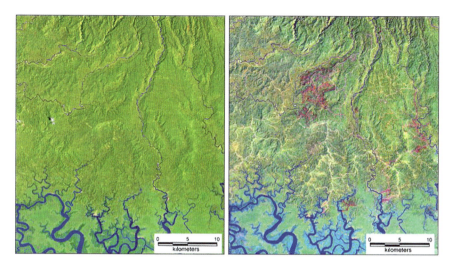

Fig. 4.8 Satellite images depicting the impact of logging in Papua New Guinea's Gulf Province. The 1988 image (*left*) shows intact rain forest, while the 2002 image (*right*) shows the impact of logging that began in 1995. Images courtesy of the University of Papua New Guinea

(3) assists in developing policies (Bauer 1975). While spatial data has been used to increase productivity and yield in croplands for some time, new approaches are being developed for specialty products and livestock.

Vineyards in France have been benefiting from a system provided by Oenoview®, jointly developed by Institut Coopératif du Vin (Groupe ICV) and Infoterra, a subsidiary of EADS Astrium. Using aerial and multispectral images (covering nearly 3,000 square kilometers) taken at the rate of 1,000 plots every 8 s, Oenoview has a precision range of approximately 3 square meters. The images are used to map the surface area and variation of leaf canopy across vineyards—a vital statistic to pinpoint vineyard vigor, soil moisture levels, bunch and grape weight, and the presence of minerals (Samuel 2011). Wine growers provide information on their fields, vines, and grape varieties. SPOT-5 (Fig. 4.9) and Formosat-2 images are taken in July and August during the period of *véraison* (when grapes begin to ripen, about three to four weeks before grape harvest). Red wine grapes require a certain amount of water stress to produce top quality wine. Spectrally, green areas of dense foliage reveals excess soil moisture, while red to purplish areas indicate less moisture, thus pointing to a good vintage. Spatial data can be supplied in GIS-ready format to be used by the growers to determine when grapes are mature for harvesting and assess variations in grape quality within plots. Rather than acquire plot information empirically over several decades, the data allows growers to separate out lower-grade bunches for *vin de table* and reserve the superior grapes for higher end *grands crus*. Vats containing as little as 20 percent immature grapes lowers the quality of wine at the end of the process, resulting in lower economic returns (Samuel 2011). The data also allows growers to optimize pruning, fertilizer inputs, and irrigation needs for each vineyard.

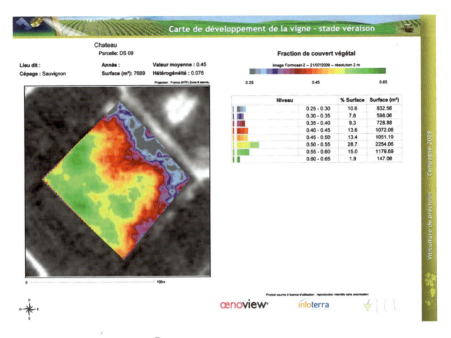

Fig. 4.9 Example of Oenoview® image http://www.astrium-geo.com

Humanitarian and Peace-Keeping Operations

It would be utterly impossible to overstate the usefulness of high-resolution satellite imagery in providing near real-time human rights-related documentation, systematically monitoring potential hotspots and threats, and advocacy efforts. High resolution imagery is especially crucial in assessing the extent of violent conflict, environmental or social justice cases, forced displacement, or other human rights atrocities in remote, prohibited, or inaccessible locations. Commercial satellites passing overhead have the ability to detect potential threats to civilians, track the movements of displaced people, provide evidence of burned and razed villages, and verify the existence of mass graves. This allows for continuous monitoring and corroboration of on-the-ground reports with the aim of decreasing or preventing humanitarian disaster and human rights crimes in politically unstable and chronic conflict areas.

Remotely-sensed images have effectively been used in legal proceedings and diplomatic negotiations to provide compelling, visual proof and resolve current and future conflicts over highly contested territories. SPOT images were used to achieve the 1995 Dayton Peace Accord (Johnson 2001). Imagery from India's IRS-1C and U.S. IKONOS satellites were used to monitor human activity and mitigate potential conflicts in Spratly Islands in the South China Sea (Gupta and Bernstein 2001).

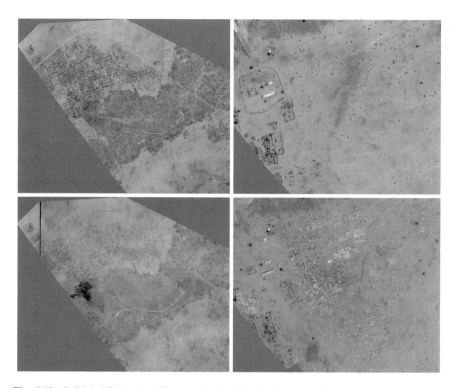

Fig. 4.10 © DigitalGlobe, Inc. Eyes on Darfur. Satellite images of Tawila, Sudan taken on 3 May 2003 (before attack, *top*), and 23 February 2007 after attacks (*bottom*), showing loss of villages and presence of Internally Displaced Persons (IDPs) camps

International humanitarian relief efforts in Thailand, Mexico, Africa, and Europe—particularly Kosovo, Macedonia, and Albania, were made possible by imagery provided by India's IRS-1C, Russia's KVR-1000, and France's SPOT satellites to monitor and mitigate the plight of refugees (Bjorgo 2001).

The well-known Eyes on Darfur project, a collaborative initiative between Amnesty International and the American Association for the Advancement of Science (AAAS) is one case study of an active monitoring project. Images allow the ability to track vulnerable villages and monitor changes that may have occurred (Fig. 4.10).

The potential of geospatial visualization for the international human rights community was highlighted by the September 2004 use of QuickBird satellite imagery by the U.S. Agency for International Development (USAID) and the U.S. Department of State in confirming the extent of ethnic cleansing in Darfur. These agencies were able to view and interpret conditions on the ground in Darfur when eyewitness reporting was difficult to verify. Used in conjunction with classified sources, analyses of the imagery led to confirmation of reports of widespread destruction of villages, livestock, and crops as part of an ethnic cleansing campaign. The remotely-sensed imagery also proved effective in diplomatic campaigns and alerting the global public.

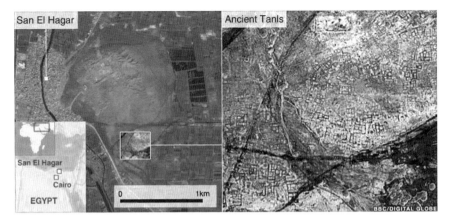

Fig. 4.11 © DigitalGlobe/BBC. The infrared image on the right used by the SS-SEPE team reveals the ancient city of Tanis near the modern city of San El Hagar

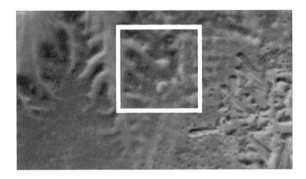

Fig. 4.12 © Sarah Parcak, SS-SEPE. The Landsat TM image reveals the outlines of a buried pyramid, located in the center of the box. Validation of the technology came when the team excavated a 3,000-year-old house that the satellite imagery had detected; the outline of the structure was almost identical to that of the remotely-sensed image

Using Remote Sensing in Archaeology

The evolution of satellite image analysis of has vastly transformed the field of archaeology, allowing researchers to exploit more fully an enormous wealth of data of the earth's surface and sub-surface contained in various types of satellite images along with aerial photography. Archaeologists can examine a broad spectrum of reflectivity signatures and bands within and between archaeological sites to determine if there are subsurface structures.

The South Sinai Survey and Excavation Project in Egypt (SS-SEPE) led by Drs. Gregory Mumford and Susan Parcak of University of Alabama, Birmingham (U.S.), used supervised and unsupervised land use/land cover applications of

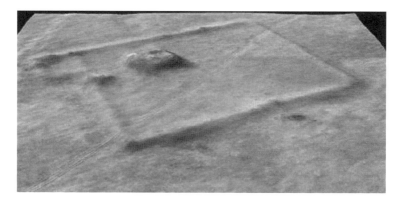

Fig. 4.13 © Jan Bemmann. Satellite images reveal the rectangular walled city of Kharbalgas, approximately 30 square kilometers

Landsat TM images as well as QuickBird imagery to detect vegetation signatures throughout Egypt. This allowed the researchers to isolate water sources in an arid environment and thus identify existing and potential archaeological sites. Infrared images can differentiate between various sub-surface materials. Since ancient Egyptians used mud bricks in building construction, which are substantially denser than the surrounding soils, the shapes of houses, temples, and tombs can be detected most easily using winter images (Pringle 2011). Sites meeting the classification parameters are then verified by "ground-truth" to confirm already identified archaeological sites (i.e., assess the accuracy of the classification), and conduct initial excavations on potential (but unknown) sites. As a consequence of creating this specialized classification, the enormous amount of time and high costs associated with surface detection of vegetation and water signatures in a vast geographical area can be significantly reduced, and the possibility for new discoveries greatly increased (Parcak and Mumford 2010).

This innovative technique rewarded the SS-SEPE team with one of the most significant discoveries in Egypt—in May 2011, 17 previously unknown pyramids along with more than 1,000 tombs and 3,000 settlements (Figs. 4.11 and 4.12) were revealed (Cronin 2011; Pringle 2011).

The vastness and remoteness of Mongolia, along with the lack of surface remains has meant that it has been largely unexplored by archaeologists. Starting in 2008, a German research team led by Dr. Jan Bemman of University of Bonn, used a combination of aerial photographs, satellite images, and surveys. Digital surface models were converted into interactive 3D models of the area and superconducting quantum interference devices (SQUIDS) on the ground were used to obtain rapid, high resolution images of the magnetic field. The combination of advanced techniques allowed for detection of compacted construction materials underneath the soil. In 2011, the team detected a massive settlement (Fig. 4.13) in the Orkhon Valley of central Mongolia dating from the eighth and ninth centuries C.E. at the time of the Uigher empire (AAAS 2011).

Safeguarding archaeological sites and antiquities has been aided by remotely sensed images. Authorities can monitor images to detect if a site has been looted and contact Interpol to watch out for antiquities that may be sold illegally.

These are just a few of the many terrestrial applications of remote sensing. The next chapter will provide examples of how remote sensing is used to better understand atmospheric phenomena.

References

S.G. Ackleson, V. Klemas, Remote-sensing of submerged aquatic vegetation in lower Chesapeake Bay: a comparison of Landsat MSS to TM imagery. Remote Sens. Environ. **22**, 235–248 (1987)

American Association for the Advancement of Science (AAAS), Beneath a barren steppe, a Mongolian surprise. New Focus. Science **332**(6028), 416–417 (2011)

R. Armstrong, Remote sensing of submerged vegetation canopies for biomass estimation. Int. J. Remote Sens. **14**, 621–627 (1993)

G.P. Asner, E.N. Broadbent, P.J.C. Oliveira, M. Keller, D.E. Knapp, J.N.M. Silva, Condition and fate of logged forests in the Brazilian Amazon. Proc. Natl. Acad. Sci. U.S.A. **103**(34), 12947–12950 (2006)

M.E. Bauer, The role of remote sensing in determining the distribution and yield of crops. Adv. Agron. **27**, 271–304 (1975)

E. Bjorgo, Supporting Humanitarian Relief Operations, in *Commercial Observation Satellites: At the Leading Edge of Global Transparency*, ed. by D. Baker, K.M. O'Connell, R.A. Williamson (Santa Monica, CA and Bethesda, MD: RAND and the American Society for Photogrammetry and Remote Sensing 2001) 408–422

British Broadcasting Corporation (BBC), BBC News: Science/Nature: Images reveal "rapid forest loss" (2008), http://news.bbc.co.uk/2/hi/science/nature/7431589.stm

P.A. Brivio, C. Giardino, E. Zilioli, Validation of satellite data for quality assurance in lake monitoring applications. Sci. Total Environ. **268**, 3–18 (2001)

S.R. Carpenter, N.F. Caraco, D.L. Correll, R.W. Howarth, A.N. Sharpley, V.H. Smith, Nonpoint pollution of surface waters with phosphorus and nitrogen. Ecol. Appl. **8**(3), 559–568 (1998)

G.P. Catts, S. Khorram, J.E. Cloern, A.W. Knight, S.D. DeGloria, Remote sensing of tidal chlorophyll-a variations in estuaries. Int. J. Remote Sens. **6**(11), 1685–1706 (1985)

F. Cronin, Egyptian pyramids found by infra-red satellite images. (BBC News, 24 May 2011), http://www.bbc.co.uk/news/world-13522957. Accessed 25 May 2011

L.M. Curran, S.N. Trigg, Sustainability science from space: quantifying forest disturbance and land-use dynamics in the Amazon. Proc. Natl. Acad. Sci. U.S.A. **103**(34), 12663–12664 (2006)

L.M. Curran, S.N. Trigg, A.K. McDonald, D. Astiani, Y.M. Hardiono, P. Siregar, I. Caniago, E. Kasischke, Lowland forest loss in protected areas of Indonesian Borneo. Science **303**, 1000–1003 (2004)

T.E. Dahl, Wetland losses in the United States 1780s to 1980s. U.S. Department of the interior, fish and wildlife service, Washington, DC (1990)

A.G. Dekker, S.W.M. Peters, The use of the thematic mapper for the analysis of eutrophic lakes: a case study in the Netherlands. Int. J. Remote Sens. **14**, 799–821 (1993)

J.S. Feierabend, J.M. Zelazny, *Status report on our nation's wetlands* (National Wildlife Federation, Washington, DC, 1987), pp. 22–46

K. Green, D. Kempka, L. Lackey, Using remote sensing to detect and monitor land-cover and land-use change. Photogramm. Eng. Remote Sens. **60**(3), 331–337 (1994)

L.S. Guild, W.B. Cohen, J.B. Kauffman, Detection of deforestation and land conversion in Rondônia, Brazil using change detection techniques. Int. J. Remote Sens. **25**(4), 731–750 (2004)

V. Gupta, A. Bernstein, Keeping an Eye on the Island: Cooperative Remote Monitoring in the South China Sea, in *Commercial Observation Satellites: At the Leading Edge of Global Transparency* ed. by D. Baker, K.M. O'Connell and R.A. Williamson (Santa Monica, CA and Bethesda, MD: RAND and the American Society for Photogrammetry and Remote Sensing, 2001) 341–345 and 353–354

M.C. Hansen, S.V. Stehman, P.V. Potapov, T.R. Loveland, J.R.G. Townshend, R.S. DeFries, K.W. Pittman, B. Arunarwati, F. Stolle, M.K. Steininger, M. Carroll, C. DiMiceli, Humid tropical forest clearing from 2000 to 2005 quantified by using multitemporal and multiresolution remotely sensed data. Proc. Natl. Acad. Sci. U.S.A. **105**(27), 9439–9444 (2008)

M.C. Hansen, S.V. Stehman, P.V. Potapov, Quantification of global gross forest cover loss. Proc. Natl. Acad. Sci. U.S.A. **107**(19), 8650–8655 (2010)

J.R. Jensen, S. Narumanlani, O. Weatherbee, H.E. Mackey Jr, Measurement of seasonal and yearly cattail and waterlily changes using multidate SPOT panchromatic data. Photogramm. Eng. Remote Sens. **52**, 31–36 (1993)

J.R. Jensen, *Remote sensing of the environment an earth resource perspective* (Prentice-Hall, Upper Saddle River, 2000), p. 544

R.G. Johnson, Supporting the Dayton Peace Talks, in *Commercial Observation Satellites: At the Leading Edge of Global Transparency*, ed. by D. Baker, K.M. O'Connell, R.A. Williamson (Santa Monica, CA and Bethesda, MD: RAND and the American Society for Photogrammetry and Remote Sensing 2001), 299, 302

S. Khorram, Development of water quality models applicable throughout the entire san francisco bay and delta. Photogramm. Eng. Remote Sens. **51**(1), 53–62 (1985)

S. Khorram, H.M. Cheshire, Remote sensing of water quality in the Neuse River Estuary, North Carolina. Photogramm. Eng. Remote Sens. **51**, 329 (1985)

S.M. Kloiber, T.H. Anderle, P.L. Brezonik, L. Olmanson, M.E. Bauer, D.A. Brown, Trophic state assessment of lakes in the Twin Cities (Minnesota, USA) region by satellite imagery. Arch. Hydrobiol. Special Issues Adv. Limnol. **55**, 137–151 (2000)

L.P. Koh, J. Miettinen, S.C. Liew, K. Ghazoul, Remotely sensed evidence of tropical peatland conversion to oil palm. Proc. Natl. Acad. Sci. U.S.A. **108**(12), 5127–5132 (2011)

T. Kuemmerle, O. Chaskovsky, J. Knorn, V.C. Radeloff, I. Kruhlov, W.S. Keeton, P. Hostert, Forest cover change and illegal logging in the Ukrainian Carpathians in the transition period from 1988 to 2007. Remote Sens. Environ. **113**, 1194–1207 (2009)

N.T. Laporte, J.A. Stabach, R. Grosch, T.S. Lin, S.J. Goetz, Expansion of industrial logging in Central Africa. Science **316**, 1451 (2007)

R.G. Lathrop, T.M. Lillesand, Utility of Thematic Mapper data to assess water quality. Photogramm. Eng. Remote Sens. **52**, 671–680 (1986)

A. Lehmann, J.B. Lachavanne, Geographic information systems and remote sensing in aquatic botany. Aquat. Bot. **58**, 195–207 (1997)

T.M. Lillesand, W.L. Johnson, R.L. Deuell, O.M. Linstrom, D.E. Meisner, Use of Landsat data to predict the trophic state of Minnesota lakes. Photogramm. Eng. Remote Sens. **49**, 219–229 (1983)

M. Linkie, R.J. Smith, N. Leader-Williams, Mapping and predicting deforestation patterns in the lowlands of Sumatra. Biodivers. Conserv. **13**, 1809–1818 (2004)

R.C. Loehr, Characteristics and comparative magnitude of nonpoint sources. J. Water Pollut. Control Fed. **46**, 1849–1872 (1974)

C.J. Madden, R.D. DeLaune, Chemistry and nutrient dynamics. the ecology of the Barataria basin, Louisiana: an estuarine profile, ed. by W.H. Conner and J.W. Day. Fish and Wildlife Service. Biological Report **85**(7.13), 18–30 (1987)

G.K. Moore, Satellite remote sensing of water turbidity. Hydrobiol Sciences **25**, 407–421 (1980)

E.F. Moran, Deforestation and land use in the Brazilian Amazon. Human Ecol. **21**(1), 1–21 (1993)

S. Narumalani, J.R. Jensen, J.D. Althausen, S.G. Burkhalter, H.E. Mackey Jr, Aquatic macrophyte modeling using GIS and logistic multiple regression. Photogramm. Eng. Remote Sens. **63**, 41–49 (1997)

National Aeronautics and Space Administration (NASA), NASA's earth observing system—EOS AM-1. Greenbelt: NASA Goddard Space Flight Center, p. 43 (1998)

National Aeronautics and Space Administration NASA, World of change: Amazon deforestation: feature articles. NASA earth observatory (2011), http://earthobservatory.nasa.gov/Features/WorldOfChange/deforestation.php

S.A.C. Nelson, K.S. Cheruvelil, P.A. Soranno, Remote sensing of freshwater macrophytes using Landsat TM and the influence of water clarity. Aquat. Bot. **85**, 289–298 (2006)

S.A.C. Nelson, P.A. Soranno, K.S. Cheruvelil, S.A. Batzli, D.L. Skole, Regional assessment of lake water clarity using satellite remote sensing 2003. J. Limnol. **62**(Suppl. 1), 27–32 (2003)

S. Parcak, G. Mumford, South Sinai survey and excavation projects in Egypt. SEPE: Satellite Imaging (2010), http://www.deltasinai.com/image-00.htm. Accessed 2 Sept 2011

M.A. Pedlowski, E.A.T. Matricardi, D. Skole, S.R. Cameron, W. Chomentowski, C. Fernandes, A. Lisboa, Conservation units: a new deforestation frontier in the Amazonian state of Rondônia. Brazil Environ. Conserv. **32**(2), 149–155 (2005)

J. Penuelas, J.A. Gamon, K.L. Griffen, C.B. Field, Assessing community type, plant biomass, pigment composition, and photosyntheic efficiency of aquatic vegetation on spectral reflectance. Remote Sens. Environ. **46**, 110–118 (1993)

G.W. Petersen, J.M. Hamlett, G.M. Baumer, D.A. Miller, R.L. Day, J.M. Russo, Evaluation of agricultural nonpoint pollution potential in Pennsylvania using a geographical information system. Final Report ME89279. Environmental Resources Research Institute, Pennsylvania State University, PA, p. 60 (1991)

H. Pringle, Satellite imagery uncovers up to 17 lost Egyptian pyramids. 27 May 2011. AAAS Science Now (2011), http://news.sciencemag.org/sciencenow/2011/05/satellite-imagery-uncovers-up-to.html?etoc&elq=c8aa9b3355d44873bf50dac382fafb07. Accessed 1 June 2011

H. Rahman, Influence of atmospheric correction on the estimation of biophysical parameters of crop canopy using satellite remote sensing. Int. J. Remote Sens. **22**(7), 1245–1268 (2001)

H. Samuel, Satellites help French winemakers pick a perfect harvest. The Telegraph (UK) (2011), http://www.telegraph.co.uk/news/worldnews/europe/france/8673485/Satellites-help-French-winemakers-pick-a-perfect-harvest.html. Accessed 1 Aug 2011

C.E. Sasser, M.D. Dozier, J.G. Gosselink, J.M. Hill, Spatial and temporal changes in Louisiana's Barataria Basin marshes, 1945–1980. Environ. Manage. **10**(5), 671–680 (1986)

P.L. Shearman, J. Ash, B. Mackey, J.E. Bryan, B. Lokes, Forest conversion and degradation in Papua New Guinea 1972–2002. Biotropica **41**(3), 379–390 (2009)

C. Song, C.E. Woodcock, K.C. Seto, M.P. Lenney, S.A. Macomber, Classification and change attention using Landsat TM data: when and how to correct atmospheric effects? Remote Sens. Environ. **75**, 230–244 (2001)

U.S. Tim, R. Jolly, Evaluating agricultural nonpoint-source pollution using integrated geographic information systems and hydrologic/water quality model. J. Environ. Qual. **23**, 25–35 (1994)

D.L. Verbyla, *Satellite remote sensing of natural resources* (CRC, New York, 1995), p. 198

D.N. Wear, T.G. Turner, R.J. Naiman, Land cover along an urban-rural gradient: implications for water quality. Ecol. Appl. **8**(3), 619–630 (1998)

Chapter 5
Using Remote Sensing
in Atmospheric Applications

Monitoring gases, radiation, water vapor, and other data using remotely-sensed images have led to an expanded comprehension of physical and chemical processes in the atmosphere over the past decade. LiDAR has been used to monitor trace gases from the ground level up to the stratosphere with a high range resolution. RADAR has been used for weather forecasting and many other tracking of weather related phenomena.

Weather Forecasting

The most common atmospheric applications of remotely sensed data include local daily weather forecasts that are reported, severe weather forecasting, and episodic events. U.S. satellites that collect data for weather forecasting are the Geostationary Operational Environmental Satellites (GOES). These satellites have 4-km spatial resolution and 30-min temporal resolution. The images viewed on television in the U.S. are from GOES satellites. The counterparts for these satellites are Meteosat in Europe and Meteor in Russia as well as weather satellites in Japan and India.

Figures 5.1 and 5.2 depict Hurricane Floyd along the Eastern United States. The true color composite was collected by SeaWIFS satellite and the classified map of the same hurricane was collected by GOES-8.

Global Climate Change

The Fourth Assessment Report from the Intergovernmental Panel on Climate Change (IPCC) asserted that warming of the Earth's climate is unequivocal, as demonstrated by observed increases in global air and ocean temperatures, large-scale melting of snow and ice, and rising sea levels (IPCC 2007a). Furthermore,

S. Khorram et al., *Remote Sensing*, SpringerBriefs in Space Development,
DOI: 10.1007/978-1-4614-3103-9_5, © Siamak Khorram 2012

Fig. 5.1 True color composite of Hurricane Floyd taken by SeaWIFS satellite in 1999. Image courtesy of NASA

Fig. 5.2 Classified image of Hurricane Floyd using GOES-8 data. The shades of *red* indicate the severity and the wind velocity, with *dark red* being the highest wind velocity. Image courtesy of NASA

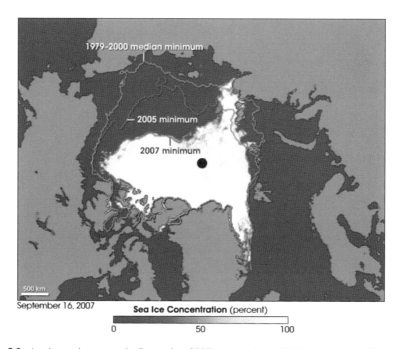

Fig. 5.3 Arctic sea ice extent in September 2007 as record by NASA's Advanced Microwave Scanning Radiometer-Earth Observing System (AMSR-E) sensor mounted on the Aqua satellite. *Blue* indicates open water, *white* indicates high sea ice concentration, and *turquoise* indicates loosely packed sea ice. Previous minimum sea ice extents are shown for comparison. The *black circle* represents a lack of data as the North Pole is beyond the maximum latitude observed by the satellite. Image courtesy of NASA and the U.S. National Snow and Ice Data Center (http://earthobservatory.nasa.gov/IOTD/view.php?id=8126)

there is substantial evidence that these changes have already impacted many physical and biological systems; for example, with respect to terrestrial ecosystems, the IPCC has very high confidence in the connection between warming and observed shifts in the ranges of certain plant and animal species (IPCC 2007a).

Remote sensing plays a prominent and ongoing role in the monitoring of phenomena associated with global climate change. For instance, in keeping with the methodology and findings of the IPPC Fourth Assessment Report, NASA's Global Climate Change website (http://climate.nasa.gov/index.cfm) provides summary data about five key quantitative indicators, all of which are currently tracked by satellite-based sensors: (1) the extent of arctic sea ice, (2) the extent of land ice, (3) sea level rise, (4) the atmospheric carbon dioxide concentration, and (5) global temperature. Below, we provide examples where specific spaceborne sensors have been applied to provide information about these five inter-related topic areas.

Arctic sea ice. In the Arctic Ocean, sea ice reaches its minimum annual extent each September. In September 2007, Arctic sea ice fell to its lowest extent ever, which was 24% lower than the previous record minimum (September 2005) and

Fig. 5.4 Envisat Advanced
Synthetic Aperture Radar
(ASAR) image of the Larsen
B ice shelf, acquired March
22, 2007. Previous historical
extents, determined using
synthetic aperture radar
(SAR) imagery from the
European Space Agency's
ERS satellite. Image courtesy
of the European Space
Agency (http://
esamultimedia.esa.int/
images/EarthObservation/
Envisat/Larsen-2007.jpg)

37% below the historical average (Comiso et al. 2008, Giles et al. 2008). Giles et al. (2008) used data from the RA-2 radar altimeter aboard the European Space Agency's Envisat satellite to look at sea ice thickness anomalies in the Arctic region. They found that average ice thickness during winter 2007–2008 was 0.26 m less than the average winter-season thickness for the previous 6 years (winter 2002–2003 to winter 2006–2007). The decline in ice thickness was particularly pronounced in the Western Arctic, where the winter 2007–2008 thickness was 0.49 m below the average of the previous 6 years.

Figure 5.3 shows an image of the Arctic region captured on 16 September 2007 by the Advanced Microwave Scanning Radiometer for EOS (AMSR-E), a sensor on NASA's Aqua satellite. The image contains three contour lines depicting annual minimum sea ice extents as reported by the U.S. National Snow and Ice Data Center. The red line is the September 2007 minimum (i.e., the record-low extent), while the green line indicates the September 2005 minimum and the yellow line indicates the median minimum extent between 1979 and 2000.

Land ice. In the context of climate change, the phrase "land ice" especially refers to the major ice sheets of Antarctica and Greenland. Temperatures on the Antarctic Peninsula have risen more than 0.1°C per decade for the past 50 years (Steig et al. 2009), while temperatures have risen sharply in Greenland since 1990 (Hanna et al. 2008). As a result of these rising temperatures, the Antarctic and

Greenland ice sheets have been losing mass at an accelerating rate. Based on data from NASA's GRACE (Gravity Recovery and Climate Experiment) satellite mission, Velicogna (2009) estimated that the mass loss of the Greenland ice sheet increased from 137 Gt per year in 2002–2003 to 286 Gt per year in 2007–2009. Similarly, in Antarctica the mass loss increased from 104 Gt per year in 2002–2006 to 246 Gt per year in 2006–2009.

On the Antarctic Peninsula, accelerating ice loss led to two major collapses of the Larsen Ice Shelf in the last two decades. First, synthetic aperture radar (SAR) images from the European Space Agency's ERS-1 satellite documented the breaking away of a 4200 km^2 section of the Larsen-A ice shelf in January 1995, which disintegrated almost completely within a few days (Rott et al. 1996). Similarly, ERS SAR images documented the rapid breakup of 2300 km^2 of the Larsen-B ice shelf during one week in March 2002, representing a large portion of the decrease in the ice shelf's total area from approximately 11,500 km^2 in 1995 to less than 3000 km^2 in 2003 (Rack and Rott 2004). Figure 5.4 is an image of the Larsen-B ice shelf, acquired in 2007 by the Advanced Synthetic Aperture Radar (ASAR) sensor onboard the European Space Agency's Envisat satellite. Previous ice extents at a series of dates between 1992 and 2002 were derived from ERS SAR or Envisat ASAR imagery. As the image suggests, the Larsen-B shelf has continued to decline steadily after its 2002 collapse.

While rising temperatures have increased surface melting of the Greenland Ice Sheet in the last couple of decades, Howat et al. (2005) argued that much of the ice sheet's rapid thinning (up to 10 m per year) may be due to recent acceleration of outlet glaciers (i.e., glaciers that flow through narrow fjords and out to sea). The researchers used RADARSAT and Advanced Spaceborne Thermal Emission and Reflection Radiometer (ASTER) imagery to look at ice flow velocities on Greenland's Helheim Glacier. They found that the outlet glacier's margin retreated landward more than 7.5 km between 2000 and 2005, as illustrated by Fig. 5.5. Because of this retreat, the mass of ice that restricted the glacier's speed was released, allowing the glacier to accelerate. Increased ice melt due to rising temperatures may lead to a feedback loop: thinning ice near the glacier's margin leads to increased calving (i.e., creation of icebergs) and further glacial retreat, releasing more ice and allowing the glacier to accelerate, which causes even more thinning (http://earthobservatory.nasa.gov/IOTD/view.php?id=6207).

Sea level rise. Sea level rise is caused by the thermal expansion of sea water due to warming temperatures, as well as melting of non-polar glaciers, ice caps, and the polar ice sheets (Vaughan 2005, IPCC 2007a; also see previous discussion). In the 20th century, the global average sea level rose approximately 1.7 mm per year (Church and White 2006); moreover, sea level rise accelerated significantly (by 0.013 ± 0.006 mm per year) during this period. In the last two decades, a series of satellite altimetry missions have facilitated global mapping of ocean surface topography, including sea surface height: TOPEX/Poseidon (launched in 1992), Jason-1 (launched in 2001), and Jason-2 (launched in 2008). The Jason-2 satellite is operated by the Ocean Surface Topography Mission (OSTM), which is a joint effort of NASA, the National Oceanic and Atmospheric Administration

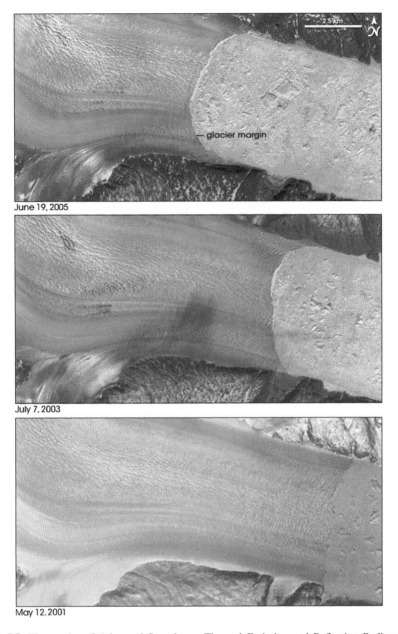

June 19, 2005

July 7, 2003

May 12. 2001

Fig. 5.5 Time series of Advanced Spaceborne Thermal Emission and Reflection Radiometer (ASTER) images of the Helheim glacier, Greenland, acquired in June 2005 (*top*), July 2003 (*middle*), and May 2001 (*bottom*). Images created by Jesse Allen, NASA Earth Observatory (http://earthobservatory.nasa.gov/IOTD/view.php?id=6207)

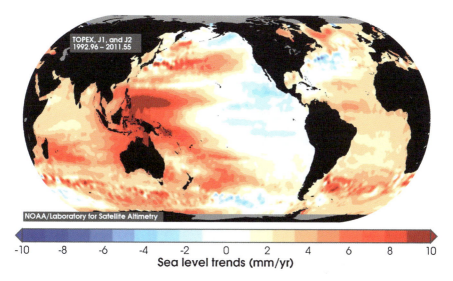

Fig. 5.6 Global map of sea level trends during the 1992–2011 time period, constructed from satellite altimetry data. Image courtesy of the U.S. National Oceanic and Atmospheric Administration (NOAA)

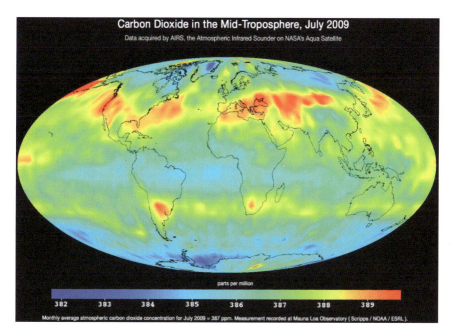

Fig. 5.7 Map of mid-troposphere CO_2 concentrations, constructed using data acquired by the Atmospheric Infrared Sounder (AIRS) sensor during July 2009. Image courtesy of NASA/Jet Propulsion Laboratory

(NOAA), the European Organisation for the Exploration of Meteorological Satellites (EUMETSAT), and France's Centre National d'Etudes Spatiales (CNES).

Figure 5.6 shows a map of sea level trends for the period between 1992 and 2011, constructed from data acquired by the TOPEX/Poseidon, Jason-1, and Jason-2 satellites. The large amount of spatial variation in sea levels reflects some influence of non-uniform changes in temperature and salinity, as well as changes in ocean circulation (IPCC 2007b). Overall, the global sea level has risen approximately 3.1 mm per year since 1993 (IPCC 2007a).

Atmospheric carbon dioxide concentration. Carbon dioxide (CO_2) is the most prominent of the "greenhouse gases" implicated in the warming of Earth's surface. Global CO_2 emissions come from natural (e.g., volcano eruptions) and anthropogenic sources; by far, the largest source of anthropogenic emissions is fossil fuel use (IPCC 2007a). Other greenhouse gases with anthropogenic sources, such as methane (CH_4) and nitrous oxide (N_2O), have also been linked to a dramatic increase in global temperature during the industrial era, and particularly in the last 50 years (IPCC 2007a).

A principal tool in current greenhouse gas monitoring and research is the Atmospheric Infrared Sounder (AIRS) instrument, which is mounted on NASA's Aqua satellite. The AIRS sensor was designed to measure the levels of atmospheric gases that influence global climate, including CO_2, carbon monoxide, methane, and ozone; it also provides profiles of atmospheric and surface (i.e., sea- and land-surface) temperature, relatively humidity and water vapor, and various cloud properties (Aumann et al. 2003). Figure 5.7 is an image of global CO_2 concentrations, created with data acquired by the AIRS sensor during July 2009. The observed concentration patterns result from the transportation of CO_2 around the Earth by the general circulation of the atmosphere. The northern hemisphere mid-latitude jet stream serves as a northern limit of enhanced CO_2. A belt of enhanced CO_2 circles the globe in the southern hemisphere, in this case corresponding to the flow of the southern hemisphere mid-latitude jet stream. This belt is fed by biogenesis activity in South America (i.e., the release of CO_2 through the respiration and decomposition of vegetation), forest fires in South America and Central Africa, and clusters of gasification plants in South Africa and power generation plants in southeastern Australia (http://photojournal.jpl.nasa.gov/catalog/PIA12339).

Global surface temperature. One of the most reliable parameters for tracking climate change is sea surface temperature (SST). In short, heat is a major determinant of global climate, and the oceans serve a major role as massive heat reservoirs   (http://science.nasa.gov/earth-science/oceanography/physical-ocean/temperature/). Sensors that provide long-term SST data suitable for climate change detection include the Advanced Very High Resolution Radiometers (AVHRR) onboard NOAA's polar-orbiting satellites, as well as the Along-Track Scanning Radiometer (ATSR) series of instruments on board the European Space Agency's ERS-1, ERS-2, and Envisat satellites (Good et al. 2007). To cite one application example, Good et al. (2007) used AVHRR data to estimate global trends in sea surface temperature for a 20-year period (1985–2004). They found

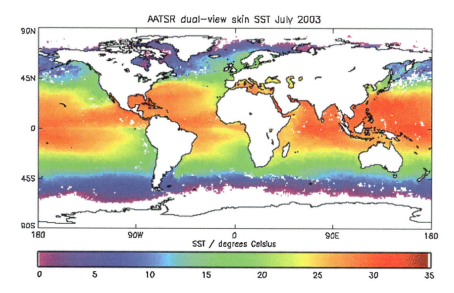

Fig. 5.8 Global map of sea surface temperature (SST) derived from data collected by the Advanced Along-Track Scanning Radiometer (AATSR) during July 2003. Image courtesy of the European Space Agency (data prepared by Anne O'Carrol, Meteorological Office, UK)

that daytime and nighttime temperatures increased 0.18° (±0.04°) and 0.17° (±0.05°) C per decade, respectively. Figure 5.8 shows an example of a global SST map derived using data, from July 2003, that were collected by the Advanced Along-Track Scanning Radiometer (AATSR) sensor onboard Envisat.

Disasters and Episodic Events

Satellite remote sensing is used in providing early warning of natural and human disasters as well as for post-disaster effects. Images gathered during and in the immediate aftermath of an earthquake, flood, fire, or any other natural or human disaster has dramatically extended the ability of international organizations, government agencies, and non-governmental organizations (NGOs) to quickly gather, analyze, and distribute reliable information. By illustrating the extent of damage and locations that are impacted, the images provide a critical tool for emergency responders and aid agencies to size up the situation and develop appropriate responses. The information guides efforts on the ground such as pinpointing still-passable roads and bridges and finding safe locations for mobile medical units or refugee shelters.

Episodic events are geologic and weather events that occur on an irregular, but periodic basis. These catastrophic events are difficult to predict, but cause overwhelming disaster to human and natural ecosystems. Episodic events include

Fig. 5.9 Massive dust storms continue to sweep across North Africa in this image acquired on 24 February 2006. Dust, which appears *grayish-pink*, is being picked up by wind from the "Great Sand Sea" region of the Sahara (and points South) and transported across the Mediterranean into Southern Europe. The verdant oasis of the Nile Delta stands in sharp contrast to the surrounding desert. Image courtesy of Jeff Schmaltz, NASA/MODIS Aqua

volcanic eruptions, earthquakes, dust storms, wildfires, and other such dramatic phenomena. This section will touch on a few of the ways remotely-sensed geo-spatial data is used.

Dust Storms and Public Health

Dust storms (Fig. 5.9) are common episodic events in dryland (deserts and semi-arid rangelands) areas around the world. In addition to severely reducing visibility and drastically impacting air quality, dust storms have deleterious effects on human health, particularly children and elderly populations, and people with respiratory or pulmonary illnesses.

For many years, public health officials thought of dust storms as merely inconvenient episodic events, however, researchers have found that dust clouds contain not only plant pollens, ashes, harmful minerals such as phosphorous, pesticide residues, industrial pollutants (e.g., heavy metals and toxic chemicals), and radioactive particles, but also living organisms—bacteria, fungus, and viruses—that transmit diseases thousands across continents to humans and livestock based on wind patterns and steering currents (Sprigg et al. 2009). The Amazon rain forest, Caribbean islands, and southeastern U.S. typically receive dust-borne particles from the African Sahara and Sahel deserts; Korea and Japan experience storms of yellow dust from China; and clouds from the Gobi and Taklimakan deserts in Asia reach the northwestern U.S. (Struck 2008). Seasonal outbreaks of meningitis and silicosis lung disease in Africa and Kazakhstan have been attributed to dust storms. Thus, if there is a pandemic such as avian flu, West Nile virus, or other disease, it can very quickly move into other areas. Prolonged droughts and climatic changes can increase the occurrence of dust storms. Scientists had originally thought that bacteria and viruses contained within dust clouds would die on exposure to ultraviolet radiation and cold temperatures, however, live microbes have been found in dust tracked as far as 6500 km (Struck 2008).

This sobering discovery prompted the U.N.'s World Meteorological Organization (WMO) to develop a satellite-based modeling system to track dust storms globally and alert those in its path. Additionally, researchers Stanley Morain of University of New Mexico (Albuquerque, New Mexico, U.S.) and William Sprigg of University of Arizona (Tucson, Arizona, U.S.) collaborated with NASA and the WMO to develop better forecasting capabilities from satellite data. Since 2004, NASA's Public Health Applications in Remote Sensing project, or PHAiRS, has worked to develop a forecasting model that takes into account wind speed and direction, near-surface temperature, surface topography, surface roughness, and the proportion of land to water using data from NASA's Terra and Aqua satellites. The resulting model has been able to predict the timing of two out of three dust storms in Phoenix, Arizona (USA).

The ultimate goal is to develop a reporting system that public health officials could use to warn the public approximately 48 h in advance and anticipate dust-related epidemics (Sprigg et al. 2009).

Managing Wildfires

Battling large-scale, catastrophic wildfires has been aided substantially by near real-time satellite data. Within the United States and Canada, the Active Fire Mapping Program is a satellite-based fire detection and monitoring program managed by the USDA Forest Service Remote Sensing Applications Center (RSAC) located in Salt Lake City, Utah, USA. This program provides near real-time detection and characterization of wildland fire conditions in a geospatial context for the United States and Canada.

Daily data collected by the NASA's MODIS satellite are currently the primary remote sensing data source of this program. The Active Fire Mapping Program leverages state of the art technologies to acquire image data directly from orbiting spacecraft in order to minimize the chance of outdated information as well as to deliver fire geospatial products as quickly as possible to incident managers, emergency response personnel and other users of this information.

Satellite image data are continually relayed to the RSAC, integrated, and processed to produce imagery and science data products. These products are processed and analyzed with current fire intelligence information and other key

Fig. 5.10 Norton Point fire along Frontier Creek. Photo: © www.inciweb.org

Fig. 5.11 Norton Point 3-D maps of the perimeter at 2:03 a.m. Mountain Time, 27 July 2011. The terrain elevation is exaggerated to make it more visible. © Google Earth/US Forest Service

geographic strata provided by U.S. and Canadian fire management agencies. The results are a suite of "value-added" geospatial products that provide an accurate and current assessment of current fire activity, fire intensity, burned area extent and smoke conditions throughout the U.S. and Canada. Products provided by the program include fire mapping and visualization products, fire detection GIS datasets and live data services, multi-spectral image subsets, and analytical products/summaries. This type of data is crucial to wildland firefighters.

The Norton Point Fire (Fig. 5.10) near Shoshone National Forest and the high country of the Washakie Wilderness in western Wyoming (US), ignited by a lightning strike during July 2011, was located in a high elevation area, making it difficult to evaluate and manage. However, within 36 h, the Rocky Mountain Type 2 Incident Management Team was able to provide detailed information to the public on a website (Fig. 5.11; see website: http://www.inciweb.org/incident/photographs/2427/).

These are just a few of the many applications for remotely-sensed atmospheric data. The next chapter presents case studies of how remote sensing has transformed our understanding of marine and planetary environments.

References

H.H. Aumann, M.T. Chahine, C. Gautier, M.D. Goldberg, E. Kalnay, L.M. McMillin, H. Revercomb, P.W. Rosenkranz, W.L. Smith, D.H. Staelin, L.L. Strow, J. Susskind, AIRS/AMSU/HSB on the Aqua Mission: design, science objectives, data products, and processing systems. IEEE. Trans. Geosci. Remote Sens. **41**(2), 253–264 (2003)

J.A. Church, N.J. White, A 20th century acceleration in global sea-level rise. Geophys. Res. Lett. **33**, L01602 (2006)

J.C. Comiso, C.L. Parkinson, R. Gersten, L. Stock, Accelerated decline in the Arctic sea ice cover. Geophys. Res. Lett. **35**, L01703 (2008)

K.A. Giles, S.W. Laxon, A.L. Ridout, Circumpolar thinning of Arctic sea ice following the 2007 record ice extent minimum. Geophys. Res. Lett. **35**, L22502 (2008)

S.A. Good, G.K. Corlett, J.J. Remedios, E.J. Noyes, D.T. Llewellyn-Jones, The global trend in sea surface temperature from 20 years of advanced very high resolution radiometer data. J. Clim. **20**, 1255–1264 (2007)

E. Hanna, P. Huybrechts, K. Steffen, J. Cappelen, R. Huff, C. Shuman, T. Irvine-Fynn, S. Wise, M. Griffiths, Increased runoff melt from the Greenland Ice Sheet: a response to global warming. J. Clim. **21**, 331–341 (2008)

I.M. Howat, I. Joughin, S. Tulaczyk, S. Gogineni, Rapid retreat and acceleration of Helheim Glacier, east Greenland. Geophys. Res. Lett. **32**, L22502 (2005)

Intergovernmental Panel on Climate Change (IPCC), Climate Change 2007: Synthesis Report. Contribution of Working Groups I, II and III to the Fourth Assessment Report of the Intergovernmental Panel on Climate Change [Core Writing Team, Pachauri, R.K and Reisinger, A. (eds.)]. IPCC, Geneva (2007a)

Intergovernmental Panel on Climate Change (IPCC), Climate Change 2007: The Physical Science Basis. Contribution of Working Group I to the Fourth Assessment Report of the Intergovernmental Panel on Climate Change [Solomon, S., Qin D., Manning M., Chen Z., Marquis M., Averyt K.B., Tignor M. and Miller H.L. (eds.)]. Cambridge University Press, Cambridge (2007b)

W. Rack, H. Rott, Pattern of retreat and disintegration of the Larsen B ice shelf, Antarctic Peninsula. Ann. Glaciol. **39**, 505–510 (2004)

H. Rott, P. Skvarca, T. Nagler, Rapid collapse of northern Larsen ice shelf, Antarctica. Science **171**, 788–792 (1996)

W. Sprigg, S. Morain, G. Pejanovic, A. Budge, W. Hudspeth, B. Barbaris, Public-health applications in remote sensing. SPIE (2009), http://spie.org/documents/Newsroom/Imported/1488/1488_5434_0_2009-02-09.pdf. Accessed 30 Oct 2011

E.J. Steig, D.P. Schneider, S.D. Rutherford, M.E. Mann, J.C. Comiso, D.T. Shindel, Warming of the Antarctic ice-sheet surface since the 1957 International Geophysical Year. Nature **457**, 459–462 (2009)

D. Struck, Dust storms overseas carry contaminants to U.S. Washington, D.C.: Washington Post. 6 February 2008

D.G. Vaughan, How does the Antarctic Sheet affect sea level rise? Science **308**, 1877–1878 (2005)

I. Velicogna, Increasing rates of ice mass loss from the Greenland and Antarctic ice sheets revealed by GRACE. Geophys. Res. Lett. **36**, L19503 (2009)

Chapter 6
Oceanographic and Planetary Applications

Remotely-sensed data has been transformative for our understanding of oceanography, marine and fisheries resources, and space exploration of other planets and extra-planetary areas. This chapter provides a few examples of how satellite remote sensing has been applied to managing vulnerable resources and gaining a better understanding of our universe.

Oceanographic Applications: Understanding Marine and Fisheries Resources

Performing environmental assessments, monitoring habitats, and developing ecosystem models in marine environments would be virtually impossible without the use of satellite remote sensing. Satellite remote sensing is also an invaluable tool for conservation ecologists and fisheries managers by providing the oceanographic information needed to understand the complexity, function, and structure of these systems and develop ecosystem approaches to sustainable fisheries management (Chassot et al. 2011). Data products derived from different wavelengths of the light spectrum include: surface optical properties such as total suspended solids, chlorophyll *a* (Fig. 6.1) and macrophytes, salinity, surface temperature (Fig. 6.2), (horizontal and vertical) circulation patterns (Fig. 6.3), pollution, sea state (waves), and current (Fig. 6.4).

Mesoscale ocean features (e.g., fronts, filaments, eddies, Lagrangian coherent structures, and river plumes), which are associated with enhanced productivity and fish aggregation, span spatio-temporal scales that range from one to thousands of kilometers and from hours to weeks. Understanding these components has been aided by the development of complex techniques and is an important aspect of marine ecosystem modeling.

Coral reefs in particular, have been extremely vulnerable to changes in ocean temperature, pollution, and algal levels. Approximately 19 percent of the world's

S. Khorram et al., *Remote Sensing*, SpringerBriefs in Space Development,
DOI: 10.1007/978-1-4614-3103-9_6, © Siamak Khorram 2012

Fig. 6.1 Chlorophyll is visible in the waters of the upwelling zone of the western South African coast in this SeaWiFS image. The sediment plume of the Orange River is also visible at the border with Namibia. Image courtesy of SeaWiFS Project, NASA/Goddard Space Flight Center, and ORBIMAGE

coral reefs are dead with another 15 percent expected to die by 2025, according to the Global Coral Reef Monitoring Network. Reefs are important as both fish habitats and buffers against high waves (Pala 2011). Australian scientists, led by Joseph Maina of Macquarie University in Sydney, conducted a global assessment of areas most susceptible to bleaching (i.e., warm water events kill corals and turn them white) by using an enormous array of satellite and tidal data. The researchers examined nearly 30 years of temperature data along with wind speed, tidal dynamics, ultraviolet radiation levels, sediment, and nutrient levels and combined that information with stress-reducing factors to identify which reefs are likely to survive additional stressors and which ones are susceptible to adverse regional conditions (Maina et al. 2011). The resulting map (Fig. 6.5) assists managers in the strategic use of conservation efforts and funds as well as policymakers.

Planetary Applications of Remote Sensing

Many of the types of sensors commonly used for remote sensing of Earth have similarly been applied to the other planets of our Solar System (Short 2010).

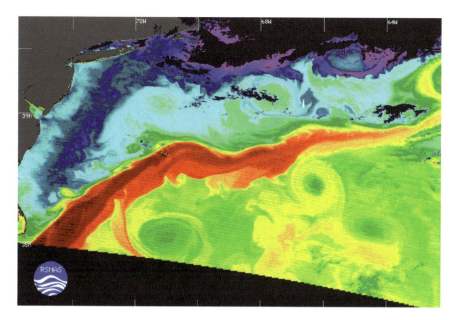

Fig. 6.2 Ocean surface temperature—Gulf Stream. Image courtesy of Bob Evans, Univ. of Miami (FL, USA) and NASA MODIS

In fact, many technological developments in remote sensing are partially rooted in planetary exploration efforts; for instance, imaging spectroscopy (i.e., hyperspectral imaging) was developed in parallel for terrestrial and planetary applications (Goetz 2009). Of course, what are perhaps the best-known planetary exploration efforts have actually involved direct (i.e., non-remote-sensing) measurements: the Apollo missions, which landed astronauts on Earth's Moon during the late 1960s and early 1970s, and more recently, the two unmanned Mars Exploration Rovers named Spirit and Opportunity, which began collecting data from the Martian surface in 2004 (Opportunity was still active as of October 2011; see the Mars Exploration Rover mission page, http://marsrover.nasa.gov/home/index.html). Notably, both of these surface investigation missions coincided with remote sensing by orbiting spacecraft. For example, the Apollo Command and Service Modules, which orbited the Moon while their corresponding Lunar Modules were on the surface, were equipped with a variety of remote sensing instruments (Short 2010). Ultimately, orbiting or fly-by spacecraft have collected remotely sensed data, in varying amounts, for all of the planets in our Solar System and some of their moons (Hanel et al. 2003), as well as asteroids and the Sun (i.e., NASA's Solar Dynamics Observatory). Here, we highlight some prominent historical missions, as well as some currently active efforts.

Unsurprisingly given its proximity, the Moon was the first target of non-terrestrial remote sensing studies. Prior to the Apollo missions, the former Soviet Union sent a series of spacecraft to the Moon, including Luna 3, which

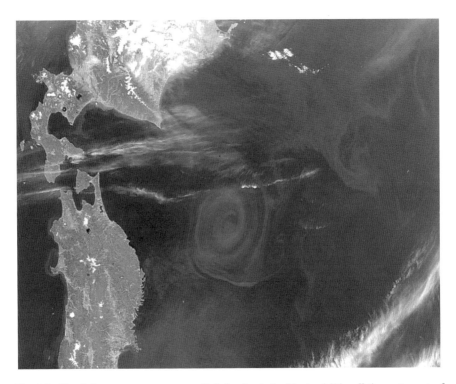

Fig. 6.3 Circulation patterns—a very well-defined spiral eddy is visible off the east coast of Japan in this SeaWiFS image. Image courtesy of SeaWiFS Project, NASA/Goddard Space Flight Center, and ORBIMAGE

captured the first images of the far side of the Moon in 1959 (Short 2010). NASA's Lunar Orbiter program similarly sent five spacecraft to the Moon between 1966 and 1967, with the primary objective to find candidate landing sites for the Apollo Lunar Modules (Short 2010).

After the final Apollo mission (Apollo 17) in 1972, the next major lunar exploration mission was the 1994 Clementine spacecraft, built and operated by the U.S. Naval Research Laboratory (McEwen and Robinson 1997). The main accomplishment of the Clementine mission was the acquisition of global-scale compositional and topographic data for the Moon. The instruments on the Clementine spacecraft included ultraviolet–visible, near infrared, long-wave infrared, and high-resolution cameras, a laser altimeter (i.e., LiDAR) system, and a radar-like unit transmitting in the S-band radio frequency (McEwen and Robinson 1997; Short 2010). Specialized products from the Clementine mission included distribution maps of compositional elements determined from recorded spectral reflectance at certain wavelengths; Fig. 6.6 is a map of the Moon's iron (i.e., reduced iron, FeO) distribution. Experiments performed with Clementine's S-band transmitter were also the first to suggest the possibility of water, in the form of ice, in the Moon's polar regions (Nozette et al. 1996).

Fig. 6.4 View of the north edge of the Gulf Stream current as it separates from the coast at Cape Hatteras, North Carolina in this NASA SeaWiFS image

NASA's Lunar Prospector, launched in January 1998 and operated until 1999 (when the spacecraft intentionally impacted the lunar surface), had a goal to further investigate Clementine's apparent finding of polar ice on the Moon; to accomplish this, its instrumentation was designed to obtain high-resolution gravity, magnetic and compositional data (Binder 1998, Konopliv et al. 2001). The instruments onboard the Lunar Prospector included gamma ray and neutron spectrometers to create global maps of the concentrations of elements such as iron, potassium, and hydrogen; a magnetometer/electron reflectometer to measure low-intensity magnetic fields; and a Doppler gravity experiment to derive a lunar gravity field map (Binder 1998). Figure 6.7 shows a gravity map created from Lunar Prospector data (Konopliv et al. 2001). The red areas depict gravitational anomalies arising from mass concentrations beneath the lunar maria, which are dark basaltic plains formed by ancient volcanic eruptions (NASA 2011).

Although the U.S. and, previously, the Soviet space programs historically dominated lunar exploration, a number of other space agencies have recently sent spacecraft to the Moon. For example, the European Space Agency launched its first lunar mission, SMART-1, in 2003; the spacecraft orbited the Moon from November 2004 to September 2006 (Foing et al. 2006). More recently, other remote sensing spacecraft have been sent to the Moon by China (Change-1 in

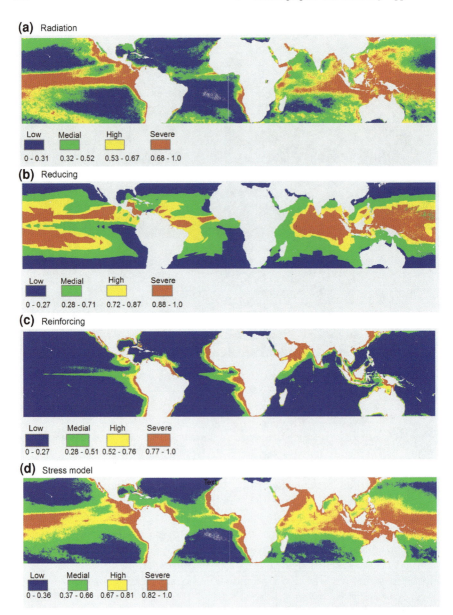

Fig. 6.5 Composite layers for radiation, reducing, reinforcing stress categories, and the overall coral reef stress model. © Maina et al. 2011

2007), Japan (Kaguya in 2008), and India (Chandrayaan-1 in 2008). The Chandrayaan-1 spacecraft, which orbited the Moon from November 2008 until August 2009, had instruments from several nations; in particular, a Brown University and NASA Jet Propulsion Laboratory research team contributed the

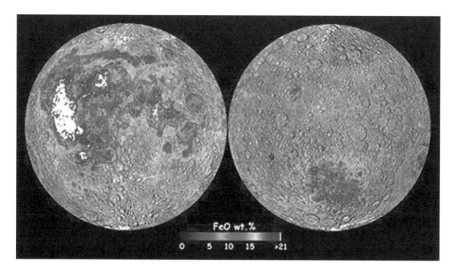

Fig. 6.6 Map of the percentage of reduced iron (FeO) on the Moon's surface, created using reflectance data from the Clementine mission. The FeO data, from 70°S to 70°N, overlay a shaded topographic map. Image courtesy of Jeffrey Gillis-Davis, University of Hawaii

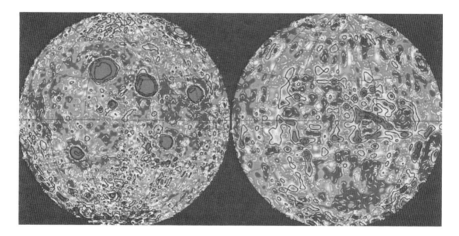

Fig. 6.7 Gravity map of the Moon made by lunar prospector in 1998–1999. Image courtesy of A.S. Konopliv, NASA. http://solarsystem.nasa.gov/yss/display.cfm?Year=2011&Month=9 (originally in Konopliv et al. 2001. Icarus 150:1–18)

Moon Mineralogy Mapper, an imaging spectrometer for collecting data at visible and near-infrared wavelengths. Significantly, measurements by the Moon Mineralogy Mapper indicated the presence of very low levels of water over most of the lunar surface (Pieters et al. 2009). Figure 6.8 shows a young lunar crater on the far side of the Moon as captured by the Moon Mineralogy Mapper. The left-hand image shows brightness at shorter infrared wavelengths, while the right-hand

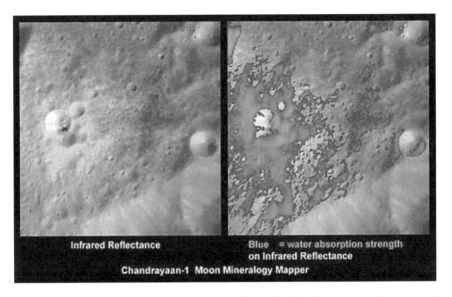

Fig. 6.8 Chandrayaan-1 Moon mineralogy mapper image showing the presence of water around a young lunar crater. Image courtesy of the Indian Space Research Organization, NASA, JPL-Caltech, U.S. Geological Survey, and Brown University

overlay image shows the distribution of water-rich minerals (light blue) in the material ejected from the crater.

The most recent Moon exploration mission is NASA's Lunar Reconnaissance Orbiter (LRO). The LRO was designed to expand upon previous lunar missions by collecting surface morphological data of sufficient resolution to support future manned exploration, including the possibility of a permanent outpost on the Moon (Robinson et al. 2010). Launched in 2009, the LRO contains a variety of instruments similar to those on previous lunar spacecraft. However, its most distinctive instrument is the Lunar Reconnaissance Orbiter Camera (LROC). The LROC consists of three imaging subsystems: two Narrow Angle Cameras for capturing high-resolution (0.5 m) panchromatic images, and a multispectral Wide Angle Camera, which captures 75-m resolution images in the visible and 384-m images at ultraviolet wavelengths (Robinson et al. 2010). In September 2011, NASA released an LROC image of the Apollo 17 landing site (Fig. 6.9). The image clearly shows foot tracks from the mission astronauts and double tracks from the lunar roving vehicle (LRV), as well as the "Challenger" Lunar Module. (Also visible are the deployment sites of the Apollo Lunar Surface Experiments Package, or ALSEP, and a set of geophones for monitoring seismic activity.)

Besides Earth, the other "rocky planets" of our Solar System (i.e., Mercury, Venus, and Mars) have been the targets of a number of historical and current remote sensing efforts.

The planet Venus was visited by 25 spacecraft during the twentieth century (including NASA'S 1990–1994 Magellan mission), but the European Space

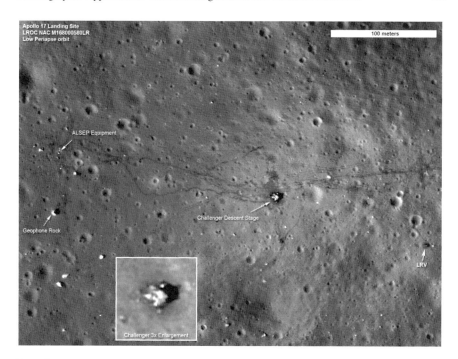

Fig. 6.9 Lunar reconnaissance orbiter camera image of the Apollo 17 landing site. Image courtesy of NASA (http://www.nasa.gov/images/content/584392main_M168000580LR_ap17_area.jpg)

Agency's Venus Express spacecraft, inserted into orbit in 2006, was the first to suggest key details about the planet's turbulent atmosphere, including the occurrence of lightning at a rate at least half that of Earth (Titov et al. 2009). Figure 6.10 is an image combining a global temperature map (lower left) of the planet's cloud tops—captured with the Visible and Infrared Imaging Spectrometer (VIRTIS) onboard Venus Express—with an ultraviolet image (upper right) captured by the spacecraft's Venus Monitoring Camera (VMC), which depicts the planet's dynamic cloud structure. Overall, the mission's findings suggest that Venus and Earth once had similar surface environments but evolved quite differently; for instance, Venus, now relatively dry, likely once held much more water but lost it because of high temperatures due to its proximity to the Sun (Titov et al. 2009).

A number of remote sensing efforts have been aimed at finding and characterizing evidence of water on Mars, either currently or historically, as a key condition for supporting life. This was a major goal of NASA's Mars Global Surveyor, which orbited the planet from 1997 to 2006 with the additional goals of studying its climate and geology and preparing for possible human exploration (NASA 2010). Among the instruments carried by the spacecraft was the Mars Orbiter Camera (MOC), comprised of 1.5-m narrow-angle and 240-m wide-angle scanners. Data collected by the MOC suggested the presence of sources of liquid water at

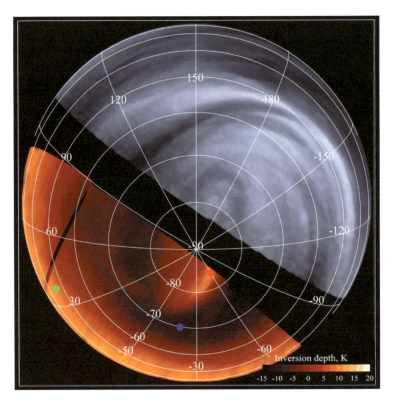

Fig. 6.10 Venus in the infrared (*lower left*) and ultraviolet (*upper right*), recorded by the Venus Express orbiter. Image courtesy of the European Space Agency

shallow surface depths (Malin and Edgett 2000); ultimately, MOC data were used to suggest that liquid water had flowed on Mars as recently as the previous decade (Malin et al. 2006). Meanwhile, another instrument onboard the Mars Global Surveyor, the Mars Orbiter Laser Altimeter, facilitated comprehensive topographic mapping of Mars (Fig. 6.11), providing 600 million measurements—with a vertical precision of better than 1 m—of the Martian surface before the sensor failed in 2001 (Kirk 2005; Smith et al. 1999).

NASA built upon the success of the Mars Global Surveyor with the Mars Odyssey spacecraft, which began mapping the planet in 2002; the spacecraft was still active in 2011. One of the primary instruments on Mars Odyssey, the Gamma-Ray Spectrometer, provided the first evidence of large subsurface water ice deposits near the planet's north and south poles (Boynton et al. 2002). The European Space Agency's Mars Express orbiter, which arrived in December 2003, further expanded on the NASA orbiter missions. In particular, its Observatoire pour la Mineʹralogie, l'Eau, les Glaces, et l'Activite (OMEGA) instrument, which is a hyperspectral visible and near-infrared imaging spectrometer, has identified some surface minerals formed in the presence of water; this suggests that the

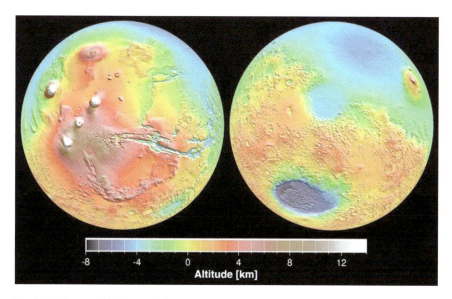

Fig. 6.11 Topographic images of the two Martian hemispheres captured by the Mars Orbiting Laser Altimeter onboard the Mars Global Surveyor. Image courtesy of NASA Jet Propulsion Laboratory

Fig. 6.12 Image of a possible dust-covered frozen sea on Mars, captured by the High-Resolution Stereo Camera on board the Mars Express orbiter. Image courtesy of the European Space Agency/DLR/Freie Universitaet Berlin (G. Neukum)

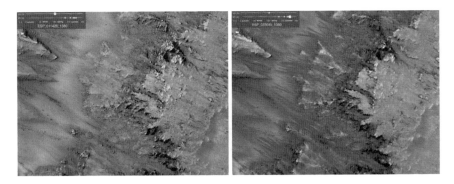

Fig. 6.13 Cold-season (*left*) and warm-season (*right*) images of the same geographic area, as captured by the Mars Reconnaissance Orbiter. Note the dark streaks extending down-slope in the warm-season image. Images courtesy of NASA

planet was once much warmer and wetter than it is now (Bibring et al. 2005, Paige 2005). Based on imagery from the High-Resolution Stereo Camera (HRSC), another instrument on board Mars Express, a team of researchers presented evidence for volcanic eruptions as recently as two million years ago on Olympus Mons, Mars' highest volcano, suggesting that the planet may still be volcanically active (Neukum et al. 2004). The study also highlighted evidence of glacial activity on Olympus Mons as recently as four million years ago. Figure 6.12 is an HRSC image showing a perspective view of a possible dust-covered frozen sea near the Martian equator. Notably, many of the fractured, plate-like features visible in this image resemble pack-ice (Murray et al. 2005).

NASA's Mars Reconnaissance Orbiter began collecting data from Martian orbit in 2006. One of the mission's primary objectives, as with its predecessors, is to study the historical and current role of water on Mars (Zurek and Smrekar 2007). The spacecraft's High Resolution Imaging Science Experiment (HiRISE) provides images with a spatial resolution as fine as 30 cm. In August 2011, scientists reported that repeat images from the HiRISE sensor reveal narrow, dark markings on steep slopes that appear to grow incrementally during warm seasons and fade in cold seasons (see Fig. 6.13); it is possible that these dark streaks represent warm-season flows of briny water (McEwen et al. 2011).

The "gas giant" planets of our Solar System (i.e., Jupiter, Saturn, Uranus, and Neptune) have been the subjects of far fewer exploratory missions than the rocky planets. NASA's Galileo spacecraft, which orbited Jupiter from 1995 to 2003, was the first spacecraft to orbit (rather than fly by, like the earlier Pioneer or Voyager missions) any of the gas giants. Galileo carried a variety of imaging systems, including near-infrared and ultraviolet spectrometers as well as a magnetometer, with which it captured detailed imagery of Jupiter's vertical cloud structure (Banfield et al. 1998) and prominent atmospheric features such as the Great Red Spot (Vasavada et al. 1998). Galileo also observed some of the planet's major moons, recording evidence of a global ocean beneath the icy surface of Europa

Fig. 6.14 Cassini image of vertical structures rising from the edge of Saturn's B ring and casting shadows on the ring. Image courtesy of NASA/JPL/Science Institute

(and possibly Ganymede and Callisto) as well as a very high level of volcanic activity on Io (McEwen et al. 1998). Another NASA spacecraft, Juno, is expected to reach Jupiter in 2016.

NASA's Cassini is the first spacecraft to orbit Saturn. Placed into orbit in 2004 and still active in 2011, Cassini has provided extensive observations of the planet's atmosphere, rings, and magnetosphere (Gombosi and Ingersoll 2010). Saturn's rings, which are largely composed of water ice, are structurally diverse and dynamic; indeed, certain structural aspects can change within the time frame of a few days (Cuzzi et al. 2010). The rings are also influenced by small (1–100 km diameter) "moonlets" embedded within them. Figure 6.14 is a Cassini narrow-angle camera image showing tall vertical structures along the edge of Saturn's "B" ring. Scientists believe that moonlets orbiting in this region strongly influence the ring material moving past them, causing these peaks to form.

Cassini has also observed some of Saturn's moons, particularly Titan. In fact, Cassini served as the launch pad for the European Space Agency's Huygens probe, which landed on the moon's surface in 2005. Scientists believe that Titan, with its dense, hazy atmosphere, strongly resembles a primitive Earth (Kerr 2005). Figure 6.15 is a Cassini radar image of the surface of Titan that appear to show filled and partly filled lakes of liquid methane; such images represent the best evidence to date of a methanological cycle on Titan that is similar to Earth's hydrologic cycle (Stofan et al. 2007).

Fig. 6.15 Radar image of the surface of Titan showing evidence of large bodies of liquid. Intensity in this colorized image is proportional to the radar backscatter (areas of *low* backscatter are *blue*, while *high*-return areas are tan). Image courtesy NASA/JPL/USGS

NASA's Voyager 2 remains the only spacecraft to have flown by Uranus (in 1986) and Neptune (in 1989). Data collected by Voyager 2 suggest that Uranus has a hot ocean of liquid water below its thick cloud atmosphere (Moore and Henbest 1986), while Neptune is extremely windy, with clouds of methane ice suspended in a clear atmosphere above a layer of hydrogen sulfide or ammonia ices (Smith et al. 1989). NASA's New Horizons spacecraft will be the first mission to study the dwarf planet Pluto and other objects in the Solar System's remote Kuiper Belt (Stern 2008). It is expected to reach Pluto in July 2015.

In terms of remote sensing beyond our Solar System, spacecraft-mounted telescopes have been critical to the detection of hundreds of extrasolar planets, or exoplanets, orbiting stars other than our Sun. An advantage of space telescopes for this task is that they avoid atmospheric turbulence and haze that impede observations by ground-based telescopes. Regardless, detection of exoplanets remain challenging because they are light-years away and orbiting stars that are far larger and brighter by comparison (Sasselov 2008). The remote sensing techniques

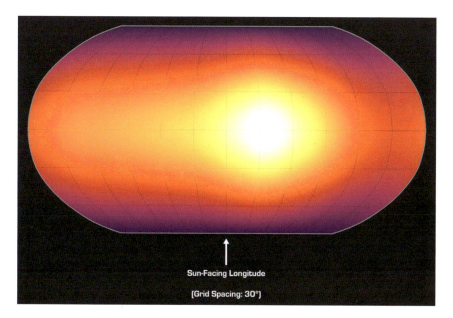

Fig. 6.16 Global temperature map for exoplanet HD 189733b, captured with the Spitzer Space Telescope. Image courtesy of NASA/JPL-Caltech/H. Knutson (Harvard-Smithsonian CfA)

involved in these analyses are simple yet provide a great deal of information. One approach is to measure the "wobble" of a star—caused by the gravitational pull of an orbiting planet—using the Doppler effect (i.e., a shift in the wavelength of the visible light from the star) (Sasselov 2008). Another method, the transiting method, looks at the drop in a star's brightness when an orbiting planet passes in front of it. Through these and other techniques, scientists are able to glean characteristics such as the planet's mass and radius, and in some cases, even more (Sasselov 2008); in 2007, a team of researchers used NASA's Spitzer Space Telescope, an infrared observatory launched in 2003, to create the first map of the surface temperature of an extrasolar planet (Knutson et al. 2007). Figure 6.16 shows the global temperature map for planet HD 189733b, a Jupiter-like gas giant. In 2008, another team of researchers reported the presence of methane in the atmosphere of HD 189733b based on observations with NASA's Hubble Space Telescope (Swain et al. 2008).

The French Space Agency (Centre national d'études spatiales), in conjunction with the European Space Agency (ESA) and other partners, launched the COnvection ROtation et Transits planétaires (CoRoT) space telescope in December 2006 (CNES 2011). NASA launched the Kepler Space Telescope in 2009 (Batalha et al. 2011). Both of these telescopes are focused on the detection of exoplanets, particularly Earth-sized planets orbiting within the "habitable zone" (i.e., the zone likely to support life) of their parent stars (Borucki et al. 2008). Each has tallied some significant accomplishments in this regard. For instance, in 2009,

CoRoT provided the first characterization of a rocky "super-Earth" planet: CoRoT-7b, which is approximately 1.58 times the size of Earth (Léger et al. 2011). In 2011, the Kepler telescope discovered its first rocky planet, Kepler-10b, which is only 1.42 times the size of Earth (Batalha et al. 2011). Regardless, future research is likely to move beyond the identification of additional rocky exoplanets to more detailed analyses of these planets' atmospheric properties, in an effort to find those with conditions most analogous to Earth (Kaltenegger et al. 2011).

References

D. Banfield, P.J. Gierasch, M. Bell, E. Ustinov, A.P. Ingersoll, A.R. Vasavada, R.A. West, M.J.S. Belton, Jupiter's cloud structure from Galileo imaging data. Icarus **135**, 230–235 (1998)

N.M. Batalha, W.J. Borucki, S.T. Bryson, L.A. Buchhave, D.A. Caldwell, J. Christensen-Dalsgaard, D. Ciardi, E.W. Dunham, F. Fressin, T.N. Gautier III, R.L. Gilliland, M.R. Haas, S.B. Howell, J.M. Jenkins, H. Kjeldsen, D.G. Koch, D.W. Latham, J.J. Lissauer, G.W. Marcy, J.F. Rowe, D.D. Sasselov, S. Seager, J.H. Steffen, G. Torres, G.S. Basri, T.M. Brown, D. Charbonneau, J. Christiansen, B. Clarke, W.D. Cochran, A. Dupree, D.C. Fabrycky, D. Fischer, E.B. Ford, J. Fortney, F.R. Girouard, M.J. Holman, J. Johnson, H. Isaacson, T.C. Klaus, P. Machalek, A.V. Moorehead, R.C. Morehead, D. Ragozzine, P. Tenenbaum, J. Twicken, S. Quinn, J. VanCleve, L.M. Walkowicz, W.F. Welsh, E. Devore, A. Gould, *Kepler*'s first rocky planet: Kepler-10b. Astrophys. J. **729**, 27 (2011)

J.P. Bibring, Y. Langevin, A. Gendrin, B. Gondet, F. Poulet, M. Berthé, A. Soufflot, R. Arvidson, N. Mangold, J. Mustard, P. Drossart, The OMEGA team of Mars surface diversity as revealed by the OMEGA/Mars express observations. Science **307**, 1576–1581 (2005)

A.B. Binder, Lunar prospector: overview. Science **281**, 1475–1476 (1998)

W. Borucki, D. Koch, N. Batalha, D. Caldwell, J. Christensen-Dalsgaard, W.D. Cochran, E. Dunham, T.N. Gautier, J. Geary, R. Gilliland, J. Jenkins, H. Kjeldsen, J.J. Lissauer, J. Rowe, KEPLER: search for Earth-size planets in the habitable zone. Proc. Intl. Astron. Union **4**, 289–299 (2008)

W.V. Boynton, W.C. Feldman, S.W. Squyres, T.H. Prettyman, J. Brückner, L.G. Evans, R.C. Reedy, R. Starr, J.R. Arnold, D.M. Drake, P.A.J. Englert, A.E. Metzger, I. Mitrofanov, J.I. Trombka, C. d'Uston, H. Wänke, O. Gasnault, D.K. Hamara, D.M. Janes, R.L. Marcialis, S. Maurice, I. Mikheeva, G.J. Taylor, R. Tokar, C. Shinohara, Distribution of hydrogen in the near surface of Mars: evidence for subsurface ice deposits. Science **297**, 81–85 (2002)

Centre national d'études spatiales (CNES) CoRoT—astronomy mission. From stars to habitable planets. CNES (2011), http://smsc.cnes.fr/COROT/

E. Chassot, S. Bonhommeau, G. Reygondeau, K. Nieto, J.J. Polovina, M. Huret, N.K. Dulvy, H. Demarcq, Satellite remote sensing for an ecosystem approach to fisheries management. ICES J. Marine Sci. 1–16 (2011)

J.N. Cuzzi, J.A. Burns, S. Charnoz, R.N. Clark, J.E. Colwell, L. Dones, L.W. Esposito, G. Filacchione, R.G. French, M.M. Hedman, S. Kempf, E.A. Marouf, C.D. Murray, P.D. Nicholson, C.C. Porco, J. Schmidt, M.R. Showalter, L.J. Spilker, J.N. Spitale, R. Srama, M. Sremčević, M.S. Tiscareno, J. Weiss, An evolving view of Saturn's dynamic rings. Science **327**, 1470–1475 (2010)

B.H. Foing, G.D. Racca, A. Marini, E. Evrard, L. Stagnaro, M. Almeida, D. Koschny, D. Frew, J. Zender, J. Heather, M. Grande, J. Huovelin, H.U. Keller, A. Nathues, J.L. Josset, A. Malkki, W. Schmidt, G. Noci, R. Birkl, L. Iess, Z. Sodnik, P. McManamon, SMART-1 mission to the Moon: status, first results and goals. Adv. Space Res. **37**, 6–13 (2006)

A.F.H. Goetz, Three decades of hyperspectral remote sensing of the Earth: A personal view. Remote Sens. Environ. **113**, S5–S16 (2009)

T.I. Gombosi, A.P. Ingersoll, Saturn: atmosphere, ionosphere, and magnetosphere. Science **327**, 1476–1479 (2010)

R.A. Hanel, B.J. Conrath, D.E. Jennings, R.E. Samuelson, *Exploration of the Solar System by Infrared Remote Sensing*, 2nd edn. (Cambridge University Press, Cambridge, 2003)

L. Kaltenegger, A. Segura, S. Mohanty, Model spectra of the first potentially habitable super-Earth—Gl581d. Astrophys. J. **733**, 35 (2011)

R.A. Kerr, Titan, once a world apart, becomes eerily familiar. Science **307**, 330–331 (2005)

R.L. Kirk, Grids & Datums: Mars. Photogramm. Eng. Remote Sens. **71**(10), 1111–1114 (2005)

H.A. Knutson, D. Charbonneau, L.E. Allen, J.J. Fortney, E. Agol, N.B. Cowan, A.P. Showman, C.S. Cooper, S.T. Megeath, A map of the day-night contrast of the extrasolar planet HD 189733b. Nature **447**, 183–186 (2007)

A.S. Konopliv, S.W. Asmar, E. Carranza, W.L. Sjogren, D.N. Yuan, Recent gravity models as a result of the Lunar Prospector mission. Icarus **150**(1), 1–18 (2001)

A. Léger, O. Grasset, B. Fegley, F. Codron, A.F. Albarede, P. Barge, R. Barnes, P. Cance, S. Carpy, F. Catalano, C. Cavarroc, O. Demangeon, S. Ferraz-Mello, P. Gabor, J.-M. Grießmeier, J. Leibacher, G. Libourel, A.S. Maurin, S.N. Raymond, D. Rouan, B. Samuel, L. Schaefer, J. Schneider, P.A. Schuller, F. Selsis, C. Sotin, The extreme physical properties of the CoRoT-7b super-Earth. Icarus **213**, 1–11 (2011)

J. Maina, T.R. McClanahan, V. Venus, M. Ateweberhan, J. Madin, Global gradients of coral exposure to environmental stresses and implications for local management. PLoS One **6**(8), e23064 (2011)

M.C. Malin, K.S. Edgett, Evidence for recent groundwater seepage and surface runoff on Mars. Science **288**, 2330–2335 (2000)

M.C. Malin, K.S. Edgett, L.V. Posiolova, S.M. McColley, E.Z. Noe Dobrea, Present-day impact cratering rate and contemporary gully activity on Mars. Science **314**, 1573–1577 (2006)

A.S. McEwen, L. Ojha, C.M. Dundas, S.S. Mattson, S. Byrne, J.J. Wray, S.C. Cull, S.L. Murchie, N. Thomas, V.C. Gulick, Seasonal flows on warm Martian slopes. Science **333**, 740–743 (2011)

A.S. McEwen, L. Keszthelyi, P. Geissler, D.P. Simonelli, M.H. Carr, T.V. Johnson, K.P. Klaasen, H.H. Breneman, T.J. Jones, J.M. Kaufman, K.P. Magee, D.A. Senske, M.J.S. Belton, G. Schubert, Active volcanism on Io as seen by Galileo SSI. Icarus **135**, 181–219 (1998)

A.S. McEwen, M.S. Robinson, Mapping of the Moon by clementine. Adv. Space Res. **19**(10), 1523–1533 (1997)

P. Moore, N. Henbest, Uranus: the view from Voyager. J.British Astron. Assoc. **96**, 131–137 (1986)

J.B. Murray, J.P. Muller, G. Neukum, S.C. Werner, S. van Gasselt, E. Hauber, W.J. Markiewicz, J.W. Head, B.H. Foing, D. Page, K.L. Mitchell, G. Portyankina, The HRSC co-investigator team evidence from the mars express high resolution stereo camera for a frozen sea close to Mars' equator. Nature **434**, 352–356 (2005)

National Aeronautics and Space Administration (NASA) Mars Global Surveyor: science summary. NASA (2010), http://mars.jpl.nasa.gov/mgs/science/

National Aeronautics and Space Administration (NASA), Solar system exploration presents: year of the solar system–September 2011: Gravity: it's what keeps us together. NASA (2011), http://solarsystem.nasa.gov/yss/display.cfm?Year=2011&Month=9

G. Neukum, R. Jaumann, H. Hoffmann, E. Hauber, J.W. Head, A.T. Basilevsky, B.A. Ivanov, S.C. Werner, S. van Gasselt, J.B. Murray, T. McCord, The HRSC co-investigator team. Recent and episodic volcanic and glacial activity on Mars revealed by the high resolution stereo camera. Nature **432**, 971–979 (2004)

S. Nozette, C.L. Lichtenberg, P. Spudis, R. Bonner, W. Ort, E. Malaret, M. Robinson, E.M. Shoemaker, The Clementine bistatic radar experiment. Science **274**, 1495–1498 (1996)

D.A. Paige, Ancient Mars: wet in many places. Science **307**, 1575–1576 (2005)

C. Pala, Coral reefs: winners and losers. 10 August 2011. AAAS Science Now (2011), http://
    news.sciencemag.org/sciencenow/2011/08/coral-reefs-winners-and-losers.html?ref=em&elq=
    3a3e6ccfdd8d43f9b7c976bac991b7f0. Accessed 22 Aug 2011
C.M. Pieters, J.N. Goswami, R.N. Clark, M. Annadurai, J. Boardman, B. Buratti, J.P. Combe,
    M.D. Dyar, R. Green, J.W. Head, C. Hibbitts, M. Hicks, P. Isaacson, R. Klima, G. Kramer, S.
    Kumar, E. Livo, S. Lundeen, E. Malaret, T. McCord, J. Mustard, J. Nettles, N. Petro, C.
    Runyon, M. Staid, J. Sunshine, L.A. Taylor, S. Tompkins, P. Varanasi, Character and spatial
    distribution of OH/H$_2$O on the surface of the Moon seen by M^3 on Chandrayaan-1. Science
    326, 568–572 (2009)
M.S. Robinson, S.M. Brylow, M. Tschimmel, D. Humm, S.J. Lawrence, P.C. Thomas, B.W.
    Denevi, E. Bowman-Cisneros, J. Zerr, M.A. Ravine, M.A. Caplinger, F.T. Ghaemi, J.A.
    Schaffner, M.C. Malin, P. Mahanti, A. Bartels, J. Anderson, T.N. Tran, E.M. Eliason, A.S.
    McEwen, E. Turtle, B.L. Jolliff, H. Hiesinger, Lunar reconnaissance orbiter camera (LROC)
    instrument overview. Space Sci. Rev. 150, 81–124 (2010)
D.D. Sasselov, Extrasolar planets. Nature 451, 29–31 (2008)
N.M. Short, The Remote Sensing Tutorial [web site]. National Aeronautics and Space
    Administration (NASA), Goddard Space Flight Center (2010), http://rst.gsfc.nasa.gov/
B.A. Smith, L.A. Soderblom, D. Banfield, C. Barnet, A.T. Basilevsky, R.F. Beebe, K. Bollinger,
    J.M. Boyce, A. Brahic, G.A. Briggs, R.H. Brown, C. Chyba, S.A. Collins, T. Colvin, A.F.
    Cook II, D. Crisp, S.K. Croft, D. Cruikshank, J.N. Cuzzi, G.E. Danielson, M.E. Davies, E. De
    Jong, L. Dones, D. Godfrey, J. Goguen, I. Grenier, V.R. Haemmerle, H. Hammel, C.J.
    Hansen, C.P. Helfenstein, C. Howell, G.E. Hunt, A.P. Ingersoll, T.V. Johnson, J. Kargel, R.
    Kirk, D.I. Kuehn, S. Limaye, H. Masursky, A. McEwen, D. Morrison, T. Owen, W. Owen,
    J.B. Pollack, C.C. Porco, K. Rages, P. Rogers, D. Rudy, C. Sagan, J. Schwartz, E.M.
    Shoemaker, M. Showalter, B. Sicardy, D. Simonelli, J. Spencer, L.A. Sromovsky, C. Stoker,
    R.G. Strom, V.E. Suomi, S.P. Synott, R.J. Terrile, P. Thomas, W.R. Thompson, A. Verbiscer,
    J. Veverka, Voyager 2 at Neptune: imaging science results. Science 246, 1422–1449 (1989)
D.E. Smith, M.T. Zuber, S.C. Solomon, R.J. Phillips, J.W. Head, J.B. Garvin, W.B. Banerdt, D.O.
    Muhleman, G.H. Pettengill, G.A. Neumann, F.G. Lemoine, J.B. Abshire, O. Aharonson, C.D.
    Brown, S.A. Hauck, A.B. Ivanov, P.J. McGovern, H.J. Zwally, T.C. Duxbury, The global
    topography of Mars and implications for surface evolution. Science 284, 1495–1503 (1999)
S.A. Stern, The new horizons Pluto Kuiper belt mission: an overview with historical context.
    Space Sci. Rev. 140, 3–21 (2008)
E.R. Stofan, C. Elachi, J.I. Lunine, R.D. Lorenz, B. Stiles, K.L. Mitchell, S. Ostro, L. Soderblom,
    C. Wood, H. Zebker, S. Wall, M. Janssen, R. Kirk, R. Lopes, F. Paganelli, J. Radebaugh, L.
    Wye, Y. Anderson, M. Allison, R. Boehmer, P. Callahan, P. Encrenaz, E. Flamini, G.
    Francescetti, Y. Gim, G. Hamilton, S. Hensley, W.T.K. Johnson, K. Kelleher, D. Muhleman,
    P. Paillou, G. Picardi, F. Posa, L. Roth, R. Seu, S. Shaffer, S. Vetrella, R. West, The lakes of
    Titan. Nature 445, 61–64 (2007)
M.R. Swain, G. Vasisht, G. Tinetti, The presence of methane in the atmosphere of an extrasolar
    planet. Nature 452, 329–331 (2008)
D.V. Titov, H. Svedhem, F.W. Taylor, S. Barabash, J.L. Bertaux, P. Drossart, V. Formisano, B.
    Häusler, O. Korablev, W.J. Markiewicz, D. Nevejans, M. Pätzold, G. Piccioni, J.A. Sauvaud,
    T.L. Zhang, O. Witasse, J.C. Gerard, A. Fedorov, A. Sanchez-Lavega, J. Helbert, R. Hoofs,
    Venus express: highlights of the nominal mission. Solar Syst. Res. 43(3), 185–209 (2009)
A.R. Vasavada, A.P. Ingersoll, D. Banfield, M. Bell, P.J. Gierasch, M.J.S. Belton, G.S. Orton,
    K.P. Klaasen, E. DeJong, H.H. Brenemann, T.J. Jones, J.M. Kaufman, K.P. Magee, D.A.
    Senske, Galileo imaging of Jupiter's atmosphere: the great red spot, equatorial region, and
    white ovals. Icarus 135, 265–275 (1998)
R.W. Zurek, S.E. Smrekar. An overview of the Mars Reconnaissance Orbiter (MRO) science
    mission. J. Geophys. Res. 112 E05S01 (2007)

Chapter 7
International Agreements and Policies

Earth observation satellites transcend national boundaries and geophysical space, creating transparency into activities and places that were once hidden from foreign states. This, of course, raises many issues: the ideals of cooperation, societal openness, and information-sharing along with the very real fears that spatial information could be used for sparking conflicts and other malevolent purposes.

Historically speaking—throughout the 1950s and 1960s—remote sensing was developed and implemented predominantly for military applications and weather observation, with nearly 75% of satellites employed for military reconnaissance and surveillance purposes (Keeley and Huebert 2004). The launch of Landsat-1 in 1972, for the purpose of "gathering facts about the natural resources of our planet" (Rocchio 2011), signaled the dawn of a new era by enabling access to widespread Earth observation by researchers and the public.

Since that long-ago day, the evolution of satellite image technology during your lifetime has allowed the opportunity for virtually everyone with access to a computer to download high resolution images easily and inexpensively. Nearly open access and ease of use is now commonplace and we would have a hard time imaging life without this convenience; this has opened the gateways to better scientific understanding of a wide variety of phenomena. However, with such easy access to literally *millions* of images covering the entire Earth's surface (and subsurface), there are understandable concerns about regulating the distribution and use of remotelysensed images and safeguarding against threats to national security.

This chapter covers international laws, agreements, and policies concerning the use of remotely sensed data, the foremost national laws and policies, and some closing thoughts about future policy directions.

Origin and Focus of International Space Law

It should not be surprising that international committees to govern the use of space and remotely sensed data were organized shortly after the first artificial satellite, Sputnik I, was launched by the Soviet Union in 1957.

S. Khorram et al., *Remote Sensing*, SpringerBriefs in Space Development,
DOI: 10.1007/978-1-4614-3103-9_7, © Siamak Khorram 2012

Fig. 7.1 United Nations Committee on the Peaceful Uses of Outer Space (COPUOS) meeting (http://www.oosa.unvienna.org/oosa/en/COPUOS/copuos.html)

The Cold War (and the quest to be the first in space exploration) between the United States and the Soviet Union was intense. Although the (spatial, temporal, and spectral) resolutions of early systems were coarse, the ability to launch satellites so soon after the end of World War II understandably triggered alarm about the ability of national superpowers to launch weapons from space, fears that national defense reconnaissance missions would be compromised, and geopolitical concerns about photographing foreign countries without their permission.

In 1958, the United Nations (UN) General Assembly established the **Committee on the Peaceful Uses of Outer Space** (COPUOS) as an *ad hoc* group to deal with the emerging issue of governing space exploration and the need to facilitate international cooperation. The following year, through UN Resolution 1472 (XIV), COPUOS was formally established as the only international forum for the development of international space law, including satellite remote sensing issues (Fig. 7.1).

Since its inception, COPUOS has concluded five international legal instruments and five sets of legal principles governing space-related activities. The five international treaties are:

1. The Treaty on Principles Governing the Activities of States in the Exploration and Use of Outer Space, including the Moon and Other Celestial Bodies (Outer Space Treaty);
2. The Agreement on the Rescue of Astronauts, the Return of Astronauts and the Return of Objects Launched into Outer Space (Rescue Agreement);
3. The Convention on International Liability for Damage Caused by Space Objects (Liability Convention);
4. The Convention on Registration of Objects Launched into Outer Space (Registration Convention); and
5. The Agreement Governing the Activities of States on the Moon and Other Celestial Bodies (Moon Treaty).

With regard to issues dealing specifically with remotelysensed data, international laws and legal principles generally focus on three matters of concern:

1. The **right to acquire** remotely sensed imagery/the right to launch remote sensing satellites;

2. The **right to disseminate** remotelysensed imagery without prior consent of the sensed State; and
3. The **right to obtain** remotelysensed satellite imagery from a particular State.

After extensive discussions in the COPUOS, the UN General Assembly unanimously adopted Resolution 41/65, **Principles Relating to Remote Sensing of the Earth from Outer Space**,[1] in December 1986. The 15 Principles place international customary obligations on countries and form the basis for remote sensing activities globally, regulating,and encouraging technical cooperation between sensing and sensed States. Several are particularly noteworthy:

- Remote sensing activities...shall be **carried out for the benefit and in the interests of all countries**, irrespective of their degree of economic or scientific development...Activities shall be conducted on the basis of respect for the principle of full and permanent sovereignty of all States and peoples...in accordance with international law... (Principle IV).
- In order to maximize the availability of benefits from remote sensing activities, States are encouraged, through agreements, to provide for the establishment and operation of data collection, storage, processing and interpretation facilities, in particular within the framework of arrangements wherever feasible (Principle VI).
- Remote sensing shall **promote the protection of mankind from natural disasters**. To this end, States participating in remote sensing activities that have identified processed data and analyzed information in their possession that may be useful to States affected by natural disasters, or likely to be affected by impending natural disasters, shall transmit such data and information to States concerned as promptly as possible (Principle XI).
- As soon as the primary data and the processed data concerning the territory under its jurisdiction are produced, the sensed State shall have access to them on a **non-discriminatory basis and on reasonable cost terms**. The sensed State shall also have access to the available analyzed information concerning the territory under its jurisdiction in the possession of any State participating in remote sensing activities on the same basis and terms, taking particularly into account the needs and interests of the developing countries (Principle XII).

While the Principles are not a binding source of international law *per se*, this watershed resolution is regarded by most space law scholars and legal researchers as the primary international legal document that addresses issues of remote sensing (Harris 2003; Jakhu 2004; Macauley 2005; Rao and Murthi 2006; Smith and Doldirina 2008). As such, the Principles are regarded as a codification of customary law and have acquired the "evidence of a general practice accepted as law" according to Article 38(1)(b), Statute of International Court of Justice (Gabrynowicz 1993; Harris 2003; Williams 2006; Smith and Doldirina 2008).

---

[1] United Nations Resolution 41/65. Principles Relating to Remote Sensing of the Earth from Outer Space. Adopted without a vote, 3 December 1986. Available online: http://www.oosa .unvienna.org/oosa/en/SpaceLaw/gares/html/gares_41_0065.html

In addition to the UN Principles, several other factors have contributed to a presumption of open access. Non-discriminatory access policies have been adopted by major remote sensing nations (e.g., Japan, the United States, and Canada), and the data policies of some remote sensing missions (e.g. ENVISAT, RADARSAT) specifically incorporate nondiscriminatory access (Harris 2003). At present there are at least 65 statutes worldwide that govern access to information, of which at least 50 establish a right of access to information, rather than a mere "limited right of access" to documents (Smith and Doldirina 2008; Gabrynowicz 1993). The right of access to environmental information is, in certain circumstances, guaranteed by the European Convention on Human Rights (Smith and Doldirina 2008).

The International Charter on Space and Major Disasters

In July 1999, following the UNISPACE III conference held in Vienna, Austria, the European and French space agencies (ESA and CNES) initiated the International Charter on Space and Major Disasters. This policy framework, which was fully operational by 1 November 2000, was the realization of Principle XI of UN Resolution 41/65—that is, to promote international cooperationand support from member space agencies and national/international space system operators during natural or technological disasters by providing critical information rapidly and managing the crisis and subsequent reconstruction operations.

Since 2000, the Charter has been increasingly activated to request satellite imagery for floods, wildfires, earthquakes, oil spills, and other catastrophic events. Authorized users are given a confidential phone number to request the mobilization of satellite remote sensing resources of the member agencies. A 24-hour on-duty operator takes the call, confirms the identity of the caller, and delivers the information to an emergency on-call officer. The emergency on-call officer assesses the request and the scope of the disaster with the authorized user. The Charter's operations team checks to see which of the members' satellites are due to pass over the affected area next. The satellite with the most appropriate sensors is used to collect the data. Data acquisition and delivery occurs on a rapid basis and a project manager assists the user in providing accurate disaster maps. The Charter is widely seen as a successful example of international cooperation and has been increasingly invoked in response to a wide variety of disasters (Fig. 7.2). It should be noted that while the Charter operates within the legally-binding framework of the Outer Space Treaty and the non-binding UN Principles, it uses a "best effort" approach with no legal binding; no liability is assumed for the resulting mapping products (Ito 2005) (Figs. 7.2, 7.3).

National Policies Governing Remotely-sensed Data

After the end of the Cold War in the late 1980s, geopolitical thinking faded and has increasingly been replaced with discourse on the collaborative uses of remote sensing data and transnational economic competition. However, the fact that

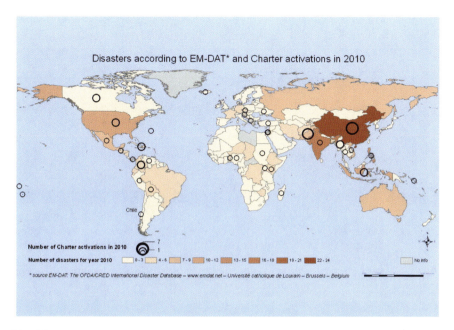

Fig. 7.2 Map of UN Charter on disasters activations during 2010. *Source*: OFDA/CRED International Disaster Database (http://www.emdat.net)

remote sensing is a dual-use technology—that is, it can be used for both military and non-military purposes—creates a tension between commercial industry interests and those of national governments.

Currently, there are approximately 30 countries with satellite-based Earth observation capabilities, compared with only three in 1980, along with an increasing number of countries which have their own image receiving stations for remote sensing systems, due to the significant reduction in acquisition costs (de Montluc 2009). Global leaders in satellite remote sensing, such as the United States, have established national laws and legal precedents, while other nations primarily rely on policies.

Common Themes and Policy Solutions

Overall, national laws and policies are rooted in similar fundamental principles-allowing access to remotelysensed imagery for scientific, social, and economic benefit, and confining access to protect national security. Before any commercial firm can design and build a remote sensing satellite, or an entity can gain access to images, they need to obtain government licenses that set or limit imaging capabilities such as panchromatic and multispectral resolutions. Additionally, once the satellites are launched, national policies also come into play in determining how

Fig. 7.3 Satellites provided images of inundated Japanese coastlines after a tsunami (*top image*) struck the nation's main island, Honshu, on March 11, 2011. Satellite images of adjacent areas, one captured on April 10, 2010 (*bottom left*) and the other on March 12, 2011(*bottom right*), illustrate the extensive destruction and flooding caused by the tsunami in the city of Ishinomaki. Images courtesy of GeoEye

remote sensing systems can be operated and the resulting images distributed. The major differences in various national policies concern high spatial and spectral resolution images (e.g., tighter restrictions, or denying access). Another method of limiting access to specific locales during specific times is known as *"shutter control"*.

BOX 7.1 What is Shutter Control?

Shutter control is a government-authorized mechanism to protect national security, international obligations, and foreign policy interests by interrupting, withholding, or blocking access to particular data. Under licensing agreements, a government could temporarily limit the imaging operations of commercial remote sensing satellites (e.g., a '24-hour rule'). An example of shutter control would be a government order (based on recommendations by national security or state department officials) to commercial remote sensing operators to turn off their satellite imaging sensors when viewing specific areas on the globe where government military operations are occurring, or are about to occur, or instances when national security could be compromised (Thompson 2007). To date, shutter control mechanisms have not been used by the U.S., but rather pseudo-shutter control, termed "persuasion to buyouts" has been used to purchase all images of a particular location to keep data on U.S. military operations from falling into the wrong hands (Gallucci 1994). Alternatives to physical shutter control include: delaying the transmission or distribution of data, restricting the field of view of the system, encryption of the data, or political/economic restrictions through contracts.

United States Laws and Policies

We begin with the United States because they were the first to develop a national law concerning remotely sensed data. Nations typically look to other nations for precedent and model their policies on this framework with differences mainly in particular national interests. Thus, other countries with specific remote sensing policies have very strong similarities to U.S. policies.

The political ideals of transparency and societal openness (in sharp contrast to the former Soviet Union's block on access) were promoted through non-discriminatory policies by installing ground stations in various regions around the world, allowing those nations non-discriminatory access to the Landsat imagery collected, and allowing the sale of the imagery (Gabrynowicz 1993). The first law concerning satellite images was passed in 1984. The Land Remote Sensing Commercialization Act (15 USC §4201) recognized that "competitive private sector participation in remote sensing was in the national interest" (Reagan 1984) and was intended to facilitate the commercialization of the Landsat system. The primary intention was to shift remote sensing away from being a government-run enterprise to private industry. This initiative was a policy failure for many reasons, including the lack of government subsidies to assist in the transition to a commercial industry, and an insufficientsatellite imagery data market.

During the early 1990s, concern about the U.S.'s prominence in global remote sensing[2] technology led to the Land Remote Sensing Policy Actof 1992 (15 USC §5601–5672). This landmark Act was initiated to ensure that the U.S. retained its international leadership role in remote sensing by promoting commercialization of unenhanced data. The aim was to maintain Landsat as "an unclassified program that operates according to the principles of open skies and non-discriminatory access" (15 USC §5601, 10). The Act transferred management of the Landsat Program from the Department of Commerce to an integrated program management by the Department of Defense and the National Aeronautics and Space Administration (NASA).

The U.S. Commercial Remote Sensing Policy, authorized by President George W. Bush on 25 April 2003,substantially relaxed restrictions on the data that commercial remote sensing industries could provide domestically and internationally. The Policy strongly encouraged and promoted the commercial remote sensing industry by shifting away from U.S. government satellite systems and to specifically, "build and operate commercial remote sensing space systems whose operational capabilities, products, and services are superior to any current or planned foreign commercial systems" (OSTP 2003). The 2003 policy guides the licensing, operation, and distribution of data and requires additional controls to protect national security and foreign policy interests. U.S. policies have created a small but growing commercial remote sensing industry and related domestic and international market for geospatial imagery. Concurrently, with the recent entry of commercial remote sensing firms into the previously military domain of satellite reconnaissance, the U.S. faces some vexing policy issues and challenges (Thompson 2007).

Legal Frameworks Within the European Union

One of the main issues confronting the European Space Agency (ESA) and memberstates of the European Union is the lack of comprehensive policies governing remote sensing data. Most European satellite operators protect data through licensing and copyright laws, however, implementation differs depending on national laws. Many individual countries control access through licensing regulations such as the Federal Republic of Germany's Satellite Data Security Act of 2007, which seeks to maximize the public's use of remotelysensed data while protecting national security. With regard to high-resolution data, Germany uses a

---

[2] France had launched SPOT in 1986 and the resulting images quickly outsold U.S. Landsat images. In 1988, India had launched IRS-1A. These events placed pressure on the U.S. to figure out how to promote the commercial remote sensing industry, which led to the passage of the Remote Sensing Policy Act of 1992. To implement that legislation, the Clinton Administration issued Presidential Decision Directive (PDD)-23 on 9 March 1994 to support and encourage foreign sales of images while protecting national security (Thompson 2007).

two-layered approach for first-time dissemination of information: a sensitivity check of the data requested by a customer and a permit/refusal by governmental authority. An algorithm determines sensitivity of the data and potential user on a case-by-case basis using factors such as whether the data are classified, the location, the customer, and other internal procedures (Schneider 2010).

Asian Policies

India developed a less restrictive Remote Sensing Data Policy, released in July 2011, which allows non-discriminatory access for images up to 1-m resolution. An inter-agency "High Resolution Image Clearance Committee" screens private and foreign users for access to sub-meter resolution imagery. The previous policy gave the National Remote Sensing Center (NRSC) a monopoly within India to restrict access to images with less than 5.8-m resolution— those from Indian and foreign satellites.

The Japan Aerospace Exploration Agency (JAXA) is an independent administrative public corporation and promotes space development and use, including remote sensing data. Japanese agencies often issue "directions", "requests", "warnings", "encouragements", and "suggestions", which are not considered to be legally binding, but rather well-respected as administrative guidance (Aoki 2010). The most recent such guidance concerning remotely-sensed data is JAXA's Basic Plan for Space Policy (2 June 2009) is the policy guidance for remotely-sensed data. In principle, all data is available to the public without limitations on spatial resolution. The policy promotes the peaceful uses of earth observation data; JAXA retains intellectual property rights.

Australian Remote Sensing Policy

Australian remote sensing policy generally concerns access control (as is the case with virtually every country's policy). Customers are required to sign a legal agreement when they order data stating that it will be used for non-commercial purposes. There are also restrictions regarding the sharing and redistribution of data. Unprocessed data cannot be transferred to a third party if it retains its original pixel structure and can be converted back to primary data. You may save a copy as a backup, but cannot otherwise duplicate the images (Geoscience Australia 2009).

Remote Sensing Policies on the African Continent

The African continent is increasingly involved in Earth observation, both as providers and users of data. Algeria, Egypt, Nigeria, and South Africa operate remote sensing satellites. The African Resources Management (ARM) Satellite Constellation is a joint program of South Africa, Nigeria, Kenya, and Algeria (and any other interested African country), aimed at fulfilling the need for regular high-resolution data over the continent for resource management applications (Dowman and Kufoniyi 2010). African countries, by and large, are members of UN COP-UOS as well as the International Telecommunication Union (ITU) and INTEL-SAT, the two intergovernmental organizations respectively responsible for the regulation and provision of communication services (Dowman and Kufoniyi 2010). However, there is a lack of national policies governing remotelysensed data. Additionally, some countries see maps and map products as highly sensitive and classified information and frequently controlled by the military. National geospatial information policies would serve to increase opportunities for capacity building, economic development, as well as innovative applications of imagery.

The Future of Remote Sensing Laws and Policy

International and national laws and policies are dynamic and ever-changing in response to changes in politics, technologies, as well as real or perceived risks to national security. While it is difficult to know what new changes will be implemented over the near future, there are some indications of particular policy directions. Internationally, two new major organizations concerned with remote sensing are the **Committee on Earth Observation Satellites** (CEOS) and the voluntary partnership of governmental and intergovernmental organizations, **Group on Earth Observations** (GEO). CEOS works toward coordinating international earth observation systems and activities to meet the common good of member states, with special attention paid to developing countries. In response to a call for more international cooperation and coordination in data sharing regarding atmospheric, land, and water data, GEO was formally created by resolution at the third Earth Observation Summit (EOS), held in Brussels in February 2005. In creating GEO on a voluntary and legally non-binding basis, the founding governments and international organizations represented resolved that GEO would implement the Global Earth Observation System of Systems (GEOSS) in accord with a 10-year (2005–2015) Implementation Plan. The Plan envisions sharing of Earth observation data by GEO members and participating organizations through the GEOSS. A major initiative of the plan is the establishment of the GEOSS Data Collection of Open Resources for Everyone (Data-CORE). Many challenges remain to be resolved, including who pays for infrastructure, training, and administration; whether to control data access; how to include the private sector; and whether problems of collective action will continue to hamper the effort.

With regard to national laws and policies, as high resolution imagery continues to be widely available through a multitude of sources, the divide between open (public) access and restricted (military) access has vanished. As a result, shutter control is not a viable national security policy; there are numerous alternative sources of high quality imagery. We have moved from an era in which a few developed countries had access to high resolution imagery to one in which virtually everyone will have such access. Transparency offers both enormous benefits and challenges. There are many complex national security concerns and policy issues that have yet to be resolved. Added to the rapid increase in the worldwide availability of high quality commercial satellite imagery, remote sensing technology and distribution has overwhelmingly outplaced the development of policy solutions. As yet, national policies globally are predominantly *ad hoc*, reactive, and not based on a working knowledge of geospatial technology. A harmonized international framework of international legal norms that goes beyond the scope of the UN Remote Sensing Principles will be needed to resolve these challenges.

References

S. Aoki, Regulation of space activities in Japan, in *National regulation of space activities*, ed. by R.S. Jakhu (Springer, New York, 2010), p. 499

B. de Montluc, The new international political and strategic context for space policies. Space Policy **25**, 20–28 (2009)

I. Dowman, O. Kufoniyi, Policies for applying earth observation in Africa: an ISPRS perspective. In: international archives of the photogrammetry, remote sensing and spatial information science, Volume XXXVIII, Part 8. Kyoto, Japan **2010**, 1088–1093 (2010)

J.I. Gabrynowicz, The promise and problems of the Land Remote Sensing Policy Act of 1992. Space Policy **9**, 319–328 (1993)

Geoscience Australia, End user license agreement. Geoscience Australia, Department of Resources, Energy, and Tourism (2009), http://www.ga.gov.au/image_cache/GA18442.pdf. Accessed 3 Oct 2011

R. Harris, Current policy issues in remote sensing: report by the International policy advisory committee of ISPRS. Space Policy **19**, 293–296 (2003)

A. Ito, Issues in the implementation of the international charter on space and major disasters. Space Policy **21**, 141–149 (2005)

R. Jakhu, International law governing the acquisition and dissemination of satellite imagery. In: J.F. Keeley, R.N. Huebert, *Commercial satellite imagery and United Nations peacekeeping: a view from above*. (Ashgate Publishing Company, Burlington, VT, 2004), p. 295f

M.K. Macauley, Is the vision of the Earth Observation Summit realizable? Space Policy **21**, 29–39 (2005)

OSTP (U.S. Office of Science Technology and Policy). U.S. Commercial Remote Sensing Policy Fact Sheet 25 April 2003 (2003), http://www.au.af.mil/au/awc/awcgate/space/2003remotesensing-ostp.htm. Accessed 11 October 2011

M. Rao, K.R.S. Murthi, Keeping up with remote sensing and GI advances—policy and legal perspectives. Space Policy **22**, 262–273 (2006)

R.W. Reagan, Statement on signing the Land Remote-Sensing Commercialization Act of 1984. 17 July 1984. *Ronald Reagan Presidential Library Archives* (1984), http://www.reagan.utexas.edu/archives/speeches/1984/71784e.htm. Accessed 25 Sept 2011

L. Rocchio, Landsat then and now. NASA (2011), http://landsat.gsfc.nasa.gov/about/. Accessed 21 Sept 2011

W. Schneider, German national data security policy for space-based earth remote sensing systems (2010), http://www.oosa.unvienna.org/pdf/pres/lsc2010/tech-02.pdf Accessed 24 Oct 2011

L.J. Smith, C. Doldirina, Remote sensing: a case for moving space data towards the public good. Space Policy **24**, 22–32 (2008)

K.P. Thompson, A political history of U.S. Commercial Remote Sensing, 1984–2007: conflict, collaboration, and the role of knowledge in the high-tech world of earth observation satellites. Doctoral dissertation, Virginia Polytechnic Institute and State University, Alexandria, Virginia, USA, 2007

M. Williams, Legal aspects of the privatization and commercialization of space activities, remote sensing, and national space legislation. 2nd Report. International Law Association, Ontario, Canada, Toronto, (2006) p. 2

Chapter 8
Future Trends in Remote Sensing

The remote sensing and associated geospatial technologies discussed in this book are just a few of the many promising areas of research and applications formulated, developed, and used by the scientific community, government agencies, nongovernmental organizations (NGOs), private industry, and the general public throughout the world. Remote sensing is a rapidly changing field and is regularly applied in a wide range of disciplines such as engineering, natural resources, geology, medicine, archaeology, anthropology, and bio-technology.

Advances in the information technology and scientific fields have and will continue to initiate and demand advances in remote sensing. More specifically, these advances will involve higher spatial, temporal, spectral, and radiometric resolution data; better image processing algorithms utilizing faster and more efficient computers and parallel processing; much larger capacity storage technology; cloud computing; rapidly evolving screen technology; and innovative applications. For instance, parallel processing strategies have rarely been implemented for remote sensing, and those that have assume homogeneity in the underlying computing architecture (e.g., a dedicated cluster of workstations with identical specifications, connected by a homogeneous communication network). However, heterogeneous distributed computer networks have emerged as a promising and cost-effective computing solution. In turn, innovative heterogeneous parallel algorithms for extracting information from high-dimensional remotely sensed data (e.g., hyperspectral imagery) are appearing on the horizon (Plaza 2006). Such advances will enable remote sensing and associated technologies to provide more focused and more accurate natural resources mapping and monitoring, quicker and more efficient emergency response, improved navigation, and better geospatial information for the general public and professionals in a wide variety of fields (Peri et al. 2001).

A somewhat unique example of an application of remote sensing, described by Ropert-Coudert and Wilson (2005), is bio-logging, in which scientists use data-recording units to acquire vast information on the behavior of animals moving freely in their natural environment. As these researchers state, this approach allows scientists to study wild animals in the field, behaving normally, with approximately

S. Khorram et al., *Remote Sensing*, SpringerBriefs in Space Development, DOI: 10.1007/978-1-4614-3103-9_8, © Siamak Khorram 2012

the same degree of rigor as laboratory studies. These systems could be applied to a wide array of aquatic and terrestrial species and allow monitoring not only the physical characteristics of the environment, but also the animal's reactions to it (Ropert-Coudert and Wilson 2005). In an era when global change and habitat losses are endangering species, this approach is critically important. Future advances involving higher data resolutions and faster data processing can assist the researchers to address such problems even more effectively.

Much higher spatial resolution data are expected to be acquired from space borne platforms in the future. One example is the hyperspectral data acquired by the Advanced Visible and Infrared Imaging Spectrometer (AVIRIS) and Hyperion with a large range of applications throughout the scientific and user communities. It is expected that these types of high-spectral-resolution sensors, with improved spatial resolution, will be regularly placed on future Earth-orbiting satellites (Vane and Goetz 1993; U.S. Congress OTA 1993). Also in this arena, many innovative approaches to refractive and reflective optics are under development. The advances in spectrometers will allow reductions in the size and mass of optical instruments. Efficient new cooling systems will extend these size benefits far into the infrared region of the electromagnetic spectrum (Pagano and Kampe 2001). There will also be innovations to realize increasingly higher sub-meter spatial resolution data, as is already being collected by GeoEye-1 (41 cm GSD), Worldview-2 (46 cm GSD), and new satellites from India, Europe, China, Japan, Russia, and others. Public–private partnerships, which have already started, will become more feasible for satellite construction and operation, and in turn, these new satellites will improve the coverage, bring down the cost, and increase the utility of remotely sensed data.

Web-based market potential for remote sensing in media has become a reality. Remote sensing data have allowed television, Internet, and print news media around the world to deliver more timely information in a more interesting way that intrigues audiences (Lurie 2011). The Internet offers the perfect medium for educating potential customers and marketing remotely-sensed data to them. News agencies, web sites, and a host of other visual media services will benefit from further advances in remote sensing and will continue to provide current, relevant, and near-real-time information regarding events around the world (Lurie 1999).

Solar power and solid-state radio frequency amplifiers will drive a new category of small remote sensing platforms, from miniaturized satellites (**"nano-satellites"**) to unmanned aerial vehicles (UAVs) and other lightweight clustered structures. Miniaturized satellites have three advantages as complements to larger systems: (1) manufacturing, launching, and operating them is cheaper; (2) they can be manufactured in a shorter time period; and (3) they have less instruments on board and are cheaper to operate. Similarly, the cost-effectiveness of UAVS could ultimately lead to a major increase in their availability to the research community and spark new remote sensing applications.

The many satellite platforms currently in low-Earth orbit (LEO) already provide us with a vast array of data, but these data are inadequate for certain applications, particularly in terms of their spatial resolution. We can expect dramatic increases in the spectral resolution of satellite-based measurements across the electromagnetic spectrum. Large antenna and mirror structures may be launched in the future to enable visible, infrared, and microwave measurements from geostationary orbits (GEO). For example, differential absorption lidar (DIAL) and laser Doppler wind measurement sensors, operated from LEO and GEO, may have apertures of several meters, permitting the collection of very high spatial resolution data (Hartley 2003). Moreover, Hartley (2003) pointed out that large antenna mirror and structures may also be placed at what are known as the Lagrange points, L1 and L2, which are located 1.5 million kilometers from Earth on the Earth–Sun line. Operating in a nearly disturbance-free environment, sensors placed at these points should be able to make highly precise measurements (Hartley 2003).

A continued decrease in the size of sensor electronics will come with further increases in processing speed. Likewise, storage density and onboard recording capacity will sustain their steady progression. By the time today's technologies are fully exploited, innovative new technologies, such as **quantum** and **biological computing**, will be realized. The ultimate result will be, for all practical purposes, unlimited computing power on orbit within the sensor itself (Hartley 2003; Cho et al. 2002).

Historically, non-governmental organizations (NGOs) and other entities who advocate for policy development (relating to environmental, social, and other issues) had limited access to remotely sensed information because the technical expertise required to process satellite imagery was restricted to government agencies. The abundance of open-access satellite imagery and image processing capabilities has resulted in broadened availability and use of remotely-sensed data outside government. This movement is further stimulated by the availability of Web-based information systems, such as Google Earth and Google Maps, which provide users easy access to satellite imagery of their backyards and beyond. These days, researchers, along with NGOs and ordinary citizens, apply remote sensing to mobilize support for situations requiring policy development—a process which has been called **satellite imagery activism** (Baker et al. 2006). Many challenges remain to be resolved, including who pays for infrastructure, training, and operation; whether and how to control data access; how to include the private sector; and whether problems of collective actions will continue to hamper the data acquisition and distribution efforts.

Remote sensing will expand to become a tool for triggering processes, particularly as it becomes more real-time or near real-time in nature. Air quality indices, for example, could possibly be monitored regionally on a daily basis, thus altering transportation policies. Once we begin to think of remote sensing in terms of more real-time (or near real-time) applications, then the scope of activities and applications will change significantly, opening the door to new possibilities and new realities for using spatial information. Data are already turned into information and will increasingly become knowledge. Remote sensing will become more valuable and we will become more reliant upon it (Thurston and Ball 2008).

In conclusion, several trends are expected to continue far into the future (Hartley 2003). These include:

- miniaturization and integration of optics and electronics;
- advances in computational power such as heterogeneous parallel computing, cloud computing, and quantum and biological computing;
- progress in large apertures and larger antennas;
- increases in acquisition and transmitter power for active systems;
- increases in storage technology and involvement of private sectors in storing and providing access to a vast amount of remotely-sensed data from the clouds;
- development of small satellites, UAVs, and lightweight structures as well as larger structures such as the International Space Station;
- advances in screen technology and mobile computing; and
- availability of hyperspectral and tunable data from satellites.

The state of the U.S. and the global economy and the burden of an increasing federal deficit in the U.S. and major European countries are forcing decision makers to seek ways to reduce the costs of remote sensing systems. This may cause major short-term challenges for the growth of remote sensing technology. These challenges will be short-lived and can best be met through the combined efforts of the international scientific and user communities.

Despite the economic pressures in development of new payloads and platforms and political challenges in data acquisition and distribution, we are enthusiastically optimistic about the future of remote sensing. In addition to the rapid growth in the technological aspects of acquiring and processing remotely sensed data, the remote sensing community is a direct beneficiary of an ever-advancing state of the art through broad collaborative efforts between developers and user communities such as biologists, engineers, social scientists, the public health and legal communities, media, urban planners, and environmental resource managers. Your opportunities to use remote sensing, in whatever career path you choose, are limitless.

References

J.C. Baker, R.A. Williamson, Satellite imagery activism: sharpening the focus on tropical deforestation. Singap. J. Trop. Geogr. **27**, 4–14 (2006)

T.-H. Chao, H. Zhou, G. Reyes, D. Dragoi, J. Hanan, High-density high-speed holographic memory, in *Proceedings of the Earth Science Technology Conference*, Pasadena, CA, 11 and 13 June 2002

J. Hartley, Earth remote sensing technologies in the twenty-first century, in *Proceedings of the International Geoscience and Remote Sensing Symposium*, Toulouse, France 1, 627–629, 21–25 July 2003

I. Lurie, Commercial future: making remote sensing a media event, in *SPIE Proceedings Vol. 3870, Sensors, Systems, and Next-Generation Satellites III*, ed. by H. Fujisada, J.B. Lurie, pp. 601–610. (1999), doi:10.1117/12.373224

J. Lurie, State-of-the-art satellite remote sensing. Opt. Photonics News **22**(1), 28–35 (2011)

T. Pagano, T. Kampe, The spaceborne infrared atmospheric sounder (SIRAS) instrument incubator program demonstration, in *Proceedings of the Earth Science Technology Conference*, College Park, MD, 28-30 Aug 2001

F. Peri Jr., J.B. Hartley, J.L. Duda, The future of instrument technology for space-based remote sensing for NASA's Earth Science Enterprise, in *Proceedings of the International Geoscience and Remote Sensing Symposium*, Sydney, Australia 1, 432–435, 9–13 July 2001

A.J. Plaza, Heterogeneous parallel computing in remote sensing applications: current trends and future perspectives. in *Proceedings of the IEEE International Conference on Cluster Computing*, Barcelona Spain, pp. 1–10, 25–28 Sept 2006

Y. Ropert-Coudert, R.P. Wilson, Trends and perspectives in animal-attached remote sensing. Front. Ecol. Environ. **3**, 437–444 (2005)

J. Thurston, M. Ball, Perspectives: what do you think the future role of remote sensing will be? Vector One Magazine, online article, posted 2 May 2008. (2008), http://www.vector1media.com/Dialogue/Perspectives/. Accessed 30 Oct 2011

U.S. Congress, Office of Technology Assessment (OTA), The future of remote sensing from space: civilian satellite systems and applications. OTA-ISC-558. Washington, DC: U.S. Government Printing Office p. 213 (1993)

G. Vane, A.F.H. Goetz, Terrestrial imaging spectrometry: current status, future trends. Remote Sens. Environ. **44**(2/3), 117–126 (1993)

Suggested Reading

I. Lurie, Commercial future: making remote sensing a media event, in *SPIE Proceedings Vol. 3870, Sensors, Systems, and Next-Generation Satellites III,* ed. by H.Fujisada, J.B. Lurie, pp. 601-610. (1999), www.portent.com/library/comfut.pdf

J. Lurie, State-of-the-art satellite remote sensing. Opt. Photonics News **22**(1), 28–35 (2011), http://www.opticsinfobase.org/abstract.cfm?URI=opn-22-1-28

Y. Ropert-Coudert, R.P. Wilson, Trends and perspectives in animal-attached remote sensing. Front. Ecol. Environ. **3**, 437–444 (2005), http://www.esajournals.org/doi/abs/10.1890/1540-9295(2005)003[0437:TAPIAR]2.0.CO;2

About the Authors

Siamak Khorram has joint appointments as a Professor of Remote Sensing and Image Processing at both the University of California (Berkeley) and North Carolina State University. He is Founder of the Center for Earth Observation, North Carolina State University. A former Vice President of Academic Programs and Dean of the International Space University (ISU) as well as a former Chair of the ISU's Academic Council. Dr. Khorram has authored over 200 publications in peer-reviewed journals and written many major technical reports. He is also a member of several professional and scientific societies. His Ph.D. is awarded jointly by the University of California at Berkeley and Davis.

Frank H. Koch Dr. Koch is a Research Ecologist with the USDA Forest Service. Previously, he was a Research Assistant Professor at North Carolina State University. His primary area of research is alien forest pest invasions. Specifically, he is interested in the spatio-temporal dynamics of invasions at national and continental scales. This multidisciplinary work involves GIS, remote sensing, statistics, and spatial simulation modeling. Dr. Koch regularly collaborates with other USDA Forest Service scientists as well as researchers from the Canadian Forest Service, the USDA Animal and Plant Health Inspection Service, and several universities. He has authored numerous journal articles and other publications. Dr. Koch received his B.A. from Duke University and his M.S. and Ph.D. from North Carolina State University.

Cynthia F. van der Wiele is currently the Director of Sustainable Communities Development, Chatham County, North Carolina. She also serves as a research associate at North Carolina State University. Her research interests include the development of high-accuracy land use/land cover classifications for analysis and improved land use planning policies. Dr. van der Wiele received her B.S. and Masters degrees from Duke University. She received her Ph.D. in community and environmental design from North Carolina State University and has performed landscape change analyses of the Research Triangle region of North Carolina.

Stacy A. C. Nelson is currently an Associate Professor and a researcher with the Center for Earth Observation at North Carolina State University. Dr. Nelson received a B.S. from Jackson State University, an M.A. from the College of William and Mary, a Ph.D. from Michigan State University. His research centers on GIS technologies to address questions of land use and aquatic systems. He has worked with several federal and state agencies including; the Stennis Space Center in Mississippi, the NASA-Regional Earth Science Applications Center (RESAC), and two Departments of Environmental Quality at the state level. He is active in several professional societies.

Index

S. Khorram et al., *Remote Sensing*, SpringerBriefs in Space Development,
DOI: 10.1007/978-1-4614-3103-9, © Siamak Khorram 2012

Printed by Printforce, the Netherlands